Strategic Management in Construction

Second Edition

David Langford
& Steven Male

Blackwell
Science

Blackwell Science Ltd
Editorial Offices:
Osney Mead, Oxford OX2 0EL
25 John Street, London WC1N 2BS
23 Ainslie Place, Edinburgh EH3 6AJ
350 Main Street, Malden
MA 02148 5018, USA
54 University Street, Carlton
Victoria 3053, Australia
10, rue Casimir Delavigne
75006 Paris, France

Other Editorial Offices:

Blackwell Wissenschafts-Verlag GmbH
Kurfürstendamm 57
10707 Berlin, Germany

Blackwell Science KK
MG Kodenmacho Building
7-10 Kodenmacho Nihombashi
Chuo-ku, Tokyo 104, Japan

Iowa State University Press
A Blackwell Science Company
2121 S. State Avenue
Ames, Iowa 50014-8300, USA

First edition published 1991 by Gower Publishing Company Ltd.
Second edition published by Blackwell Science Ltd 2001

Set in 10/12.5pt Palatino
by DP Photosetting, Aylesbury, Bucks
Printed and bound in Great Britain by
MPG Books Ltd, Bodmin, Cornwall

DISTRIBUTORS

Marston Book Services Ltd
PO Box 269
Abingdon
Oxon OX14 4YN
(*Orders:* Tel: 01235 465500
Fax: 01235 465555)

USA
Blackwell Science, Inc.
Commerce Place
350 Main Street
Malden, MA 02148 5018
(*Orders:* Tel: 800 759 6102
781 388 8250
Fax: 781 388 8255)

Canada
Login Brothers Book Company
324 Saulteaux Crescent
Winnipeg, Manitoba R3J 3T2
(*Orders:* Tel: 204 837-2987
Fax: 204 837-3116)

Australia
Blackwell Science Pty Ltd
54 University Street
Carlton, Victoria 3053
(*Orders:* Tel: 03 9347 0300
Fax: 03 9347 5001)

A catalogue record for this title is available from the British Library

ISBN 0-632-04999-5

Library of Congress
Cataloging-in-Publication Data

Langford, D.A.
Strategic management in construction/David Langford and Steven Male.—2nd ed.
p. cm.
Includes bibliographical references and index.
ISBN 0-632-04999-5
1. Construction industry—Management.
2. Strategic planning. I. Male, Steven. II. Title.

TH438 .L36 2001
624'.068—dc21

2001037773

For further information on
Blackwell Science, visit our website:
www.blackwell-science.com

Contents

Preface	*vi*
1 Introduction	**1**
Part A – Construction and its business environment	
2 Construction – an overview of the market	**11**
Industrial building	12
Commercial building	13
The housing market	14
Repairs and maintenance	17
Making sense of the trends	17
3 The strategic role of the actors in the construction process	**28**
The evolution of the construction industry in the post World War II era	28
4 Clients, constructors and competencies	**42**
Introduction	42
The concepts of product and project life cycles in construction	42
The strategic concepts of industry and market in construction	43
Market structure, price determination and competition in construction	45
Conclusion	52
Part B – Concepts of strategic management	
5 The firm and the strategic management process	**59**
Introduction	59
The strategic management process	67
The strategic management process and organisational change	80
The management of change in construction	82
Strategic behaviour	83
Conclusion	89

6 Strategic behaviour of construction firms **92**
Introduction 92
Core business and core competencies in construction 94
Levels of strategy 99
Managing the diversified construction firm 100
Strategies at the operating core in contracting firms 111
Project portfolios and potential capacity 112
Sub-contracting as a production strategy within project portfolios 113
The management resource in construction firms as a source of competitive advantage - resolving a strategic paradox 115
Conclusions 119

7 Strategies for international construction **123**
Background 123
International business strategy 123
Size and structure 125
Reasons for internationalism 129
Characteristics and obstacles to internationalisation 132
Competitive advantage and strategy in international construction 135
Competitive advantage in international construction 136
Factor conditions 141
Domestic demand conditions 142
Related and supplier/sub-contractor industries 142
Firm strategy, structure and competitiveness 142
Country analysis 144
General overview 146
Strategic positioning competitive intelligence 150

Part C - Techniques for the strategic planner

8 Portfolio management, Delphi techniques and scenarios **157**
Business portfolio management 157
The Boston Consulting Group growth–share matrix 157
Portfolio management and the construction industry 159
Delphi techniques 162
Scenarios 166
Cross impact analysis 171

9 Marketing and promotional strategies in construction **175**
Introduction 175
The marketing concept 176
Marketing orientation and relationship marketing philosophy 176
Service quality and customer satisfaction 177
Internal marketing 178
Internal customer satisfaction 179

Customer care 179
Marketing strategies - market choice and segmentation 180
The marketing mix in service industries 183
Promotional strategies 183
Promotional media (personal/non personal) 184
Co-focusing on customer service - the problems to be overcome 186

Part D - Summary

10 A synthesis of strategic management in construction 195
Introduction 195
Strategic management in the construction industry 196
Industries and markets in construction 197
Entry and exit barriers exist in an industry 198
The nature of the firm 200
The firm's product in construction 201
Core business, core competencies and distinctive capabilities 201
Innovation in construction 202
Typologies of firms 203
Strategy 204
Structuring the strategic management process 206
The decision-making role of the strategist 208
The elements of competitive strategy 210
The marketing function and its contribution to competitive advantage 211
Tools, techniques and methodologies for competitive advantage 214
Construction corporations and contracting corporations 216
Structure, strategy and process - managing the diversified construction firm 219
Construction strategies 220
Project portfolios and potential capacity 223
Sub-contracting as a production strategy within project portfolios 224
Management as a source of competitive advantage 225
Supply chain management in construction 226
International construction 229
Conclusions on managing construction corporations 230
The consultants - the construction professions 231
Implications for professional practices and the building professions by size of firm 233
Conclusions on managing construction consultancies 235
Conclusions 236
A contingency model of strategic management for construction 237

Index 243

Preface

This book is aimed at students, graduates, and practitioners in all fields of the built environment.

The second edition has been subject to a considerable overhaul from the publication of the first edition in 1991. At that time the construction industry was in the midst of a severe recession and construction corporations were actively considering strategic issues. The professions were starting to emerge as multidisciplinary practices with which clients were encouraged to have 'one stop shop' relations. Since the first edition the construction world has changed and the context of the second edition is one of an industry with a burgeoning workload and yet with very small profit margins. The time between the editions have provided the construction industry with new challenges which drive the industry – the Latham and Egan reports have spurred the industry to perform better and these pivotal reports have accelerated the evolution of the industry. As part of this renaissance of the industry, clients became more powerful and better informed, and more knowledgeable about commissioning, owning and operating their buildings. As a response constructors and the professions in the industry have had to act strategically rather than opportunistically. These changes have engendered a new vocabulary, and partnering, strategic alliances, supply chain management, lean construction and so on have all been part of the agenda of change. In this volume the authors have sought to explain the strategic importance of these issues by bringing together ideas from the world of economics, marketing, management, politics and business to serve the making of strategy in construction firms and the associated professions.

The authors have attempted to cover a number of important themes in the book. To this end it differentiates strategic management practice appropriate for contractors from professional practices, and also offers insights from theory and how this may be applied in practice. In a world shrunk by rapid communications construction organisations have also moved into being international organisations and a new chapter has been added on international strategy. This has been complemented by a chapter on marketing that has demonstrated how marketing can contribute to the strategic planning of construction organisations.

Meeting the challenges given to the construction industry will require the manager of today and tomorrow to think strategically about the future directions of their organisations. We hope that this book makes a contribution to this thinking.

Acknowledgement

The authors would like to acknowledge the contribution of many people in the production of this book. In particular the work of Marianne Halforty in producing the material from the authors', often scribbled, drafts. The work of Polly Dyne Steel in serving as our editor has made a powerful impact upon the presentation and layout of the book. For their efforts and others we are grateful.

The authors are indebted to the project team on the Glasgow Science Centre, in particular Carillion, for allowing the use of the photograph in the cover of the book.

1 Introduction

At a meeting of the International Strategic Management Society in 1996 the audience was presented with some alarming data from American business. In ten years 40% of the top 500 firms had disappeared; 60% of the top 500 firms of 1970 had been acquired or liquidated. In the Dow Jones index of 1900 only one of the top twelve companies remained in the top echelon – General Electric. The situation in Europe is largely similar. Dot.com companies temporarily displaced well established firms in the FTSE index with alarming regularity for a short space in time. In the construction industry the level of turbulence is even greater. Household names of the industry, Wimpey, Tarmac, Higgs and Hill, Trafalgar House – firms that led the way in the post-war reconstruction of Britain and indeed much of the commonwealth – have disappeared to be replaced by trans-European companies. The message of these shifts is that management in dynamic industries such as construction need to view their strategy in a more considered and structured way than they may have done in the past, but moreover agility in forming strategy with a requirement for their eyes to be on the horizon as well as the bottom line.

This strategic focus has been prompted by the sense of turbulence in the industry, with the UK Government and private clients seeking better performance of the industry. In turn the financial institutions have encouraged firms to think more strategically. One consequence of this is that firms have short term financial plans within a long term framework. This book seeks to synthesise the theoretical thinking that underpins strategic management and some of the techniques that managers may use to create strategic plans.

In order to assist the practising manager and the scholar of strategic management in construction the Second Edition has been broken down into four parts:

- Part A describes the industrial context of construction;
- Part B focuses upon fundamental concepts of strategic management;
- Part C is dedicated to recording some applications of strategic management which may be useful to the practitioner;
- Part D synthesises and summarises the ideas that have been presented in this book.

The book starts with some introductory descriptions of the construction industry and the principal markets in which it operates. It discusses how the total market for construction can be segmented into a series of overlapping sub-markets. The growth potential for each of these sub-markets is evaluated and it is noted that construction companies and consultants have increasingly specialised in one or two sub-markets. In settings where the firm is large enough to provide coverage across all sub-markets then it is likely to organise its operations such that parts of the firm are dedicated to doing particular types of work. The reasons for this are explored and may be summarised as being related to the development of particular competencies in a specific type of construction or handling particular types or sizes of projects. The analysis of the industry's markets is complemented by a glance at the structure of the industry. Despite its ups and downs the size and scale of the industry are impressive. It is a £6 billion industry providing work for $1\frac{1}{2}$ million people, producing 7% of the UK's GDP. What has been remarkable has been the extent of the repair and maintenance (R&M) market which now constitutes almost half of the workload. This huge amount of R&M goes some way to explain the structure of the industry. The large firms (over 1200 employees) are few in number and focus on the new build market – they only carry out 12% of the work. This R&M activity is the province of the small firms (1–7 employees) who populate the industry in great numbers. Small firms make up over 90% of all firms operating in the industry and large firms less than 1% of the total number. However large firms still undertake the lion's share of the construction workload.

As the industry seeks to respond to the challenges of the Egan report, pressures from government and private client to improve its time, cost, quality and safety performance, major restructuring of the industry seems set to go ahead. Transnational take-overs, domestic mergers or asset swaps, management buy-outs or Brought in Management Buy-Outs (BIMBOs) where existing managers are displaced before a management buy-out is undertaken, and the creation of one stop shop services, which include financing, designing, constructing and managing a facility, will force industry players to behave in a more proactive and strategic fashion and so replace the traditional reactive and ad hoc responses to changes in the business environment.

The changing contours of the business environment for construction are discussed in Chapter 3. The key features of the changes are related to how the markets for construction are structured, how the price for construction work is determined and the nature of competitive rivalry in the industry. This competitive character is often played out in the context of the different services that firms offer. The business environment will shape not only the kind of projects a firm chooses to chase but also the procurement and contractual choices it offers clients.

Analysing the business environment preordains a level of strategic thinking. The argument that strategic thinking in construction is not possible due to the turbulence of the construction market is challenged. Whilst it is recognised that long term strategic management is likely to be more difficult in a dynamic industry where demand is derived by the business futures of other industries the chapter argues that the implied difficulties are there to be overcome. Chapter 4, the last of those providing the industrial context carries this 'can't plan; won't plan' argument further by presenting a discussion of the generic strategies that are available to construction contractors and consultants. Strategic development for these organisations will depend upon the selected markets in which the firms seek to compete and whether the business model used by a firm is regarded as a service or a product. Design is clearly a service in which the product is largely intangible and is developed alongside the clients. In contrast speculative housebuilders are selling a product. (Although most new estates are 'signed' in such a way to sell a lifestyle – a scruffy piece of contaminated land in a less desirable part of town is given an appellation such as 'Poachers Glen', or 'Merryfield Grange' or some such to signify to the purchasers, nay 'homeowners', that they are buying a rural Arcadian lifestyle rather than a house.)

The changing role of the professions over the last twenty years is emphasised and the implications of the changing functions of architects and quantity surveyors in particular has ushered in an awareness of the necessity for strategic thinking and planning at the level of professional practice as well as the construction firm.

Part B of the book spans Chapters 5, 6 and 7 and addresses the concepts of strategy when experienced by contractors and consultants (Chapter 5), how strategy is made (Chapter 6) and firms of operating in the international arena (Chapter 7). Chapter 5 aims to explain the role of corporate strategic management and differentiates strategy making at a corporate centre from that of strategy applying to a business unit within a larger corporation. Following a deconstruction of the levels of strategy the work explores the financial factors which drive the development of strategy and the generic choices available to the strategic planner. Questions of diversification of product or service offerings are discussed along with the appropriate structure to realise a chosen strategy. Shortly put, these can be divisionalised, regionalised or based around a holding company. Finally this chapter synthesises the ideas of strategy and structure by a discussion of how strategy is realised at what is called the 'operational core' of the construction firm – the place where foundations are dug, bricks get laid, etc. – the building site.

But not only are contractors vitally interested in strategy; the consultants and professionals serving the industry have become more aware of their role as businesses. This concept of consultants as business organisations is relatively recent – certainly deregulation of the professions and the introduction of fees driven by competitive tendering sharpened the idea of consultants as businesses. Prior to deregulation most consultants earned a good living by a system of fixed fees which escalated as the costs of construction increased with inflation on fluctuating contracts. Consultants barely recognised themselves as business organisations – they were 'practices' of architects, 'partners' in the case of engineers and quantity surveyors (QSs), 'chambers' of lawyers. As changes beset the industry, these same 'practices' sought to change the way in which they did business. Chaotic responses to turbulent environments ensued.

Many sought strategic alliances with contractors for design and build work, others spread the services that they offered. QSs saw conventional bill bashing die out as new procurement methods pervaded the industry, and they reconstructed themselves as project managers. Information Technology became a driver for change; practices with long traditions of the habit of 'dotting on' became dottingon.com!

All of these changes meant that professional service companies had to embrace a business culture in which competitive advantage had to be prised by harnessing management techniques such as marketing and business planning, yet the professions have problems in presenting a unique selling point. The whole point of a profession is to even out standards of service to clients. Solving this conundrum is at the heart of strategy making for consultants.

Who will seek to position themselves in spaces in the market in which they can compete? In the design fields this can be by competing for certain types of projects and gaining a reputation for wanting to own technically challenging problems with an innovative culture pervading the professional firm. It may be by being faster or more productive by the use of IT or better at marketing, which is usually relational marketing and is seldom done by advertising.

The chapter goes on to report some pioneering work in relating the strategic groups to which the professions will belong. The largest contractors are hardly likely to be in competition with a local builder (and vice versa); professional firms can be grouped into competing clusters which may be defined by size, type of work, use of IT, etc. The discussions of the construction consultants leads on to the last subject of Part B, international strategy. Strategy in this territory will be important to contractors and consultants.

Chapter 6 seeks to explain the process of forming strategy. This part of the book sees construction as a global activity. Increasingly, the world is occupied by clients who seek to procure construction on a global basis. Neo-liberal trade regimes have promoted globalisation (and the protest against it) and the international construction industry has been greatly influenced as part of this process.

In international construction the stakeholders needing to be satisfied will expand and this imposes fresh challenges on the strategy making process. Other international projects are joint ventures by design to spread risk or by edict of the host country. This brings added complexity in that strategies of two organisations have to be melded. But of principal importance will be the analysis of the business environment of the host country. Issues such as attitude to business, security of staff and of intellectual and tangible products, cultural norms, language, the competitive environment and the challenges of the physical world will all be features of the backcloth against which the strategy for operating in an international environment needs to be set. The first strategic decision is where to seek work? Attempting to anticipate future markets is a key aspect in international construction. Often the developing world is seen as a potential market; these countries are seen to be stratifying. The poorer nations, predominantly in Africa, continue to rely on world funding for infrastructure development. World Bank policies impose a development model which insists upon private finance initiative or BOT type arrangements. In the second tier are the South American developing economics with growing political stability in the region which has enabled international investment which generates construction work but also creates the economic conditions for infrastructure development.

In Asia some economics are rapidly developing and China, in particular is a vast construction site. Other areas such as Indonesia must await calmer political waters before demand can be resurrected.

Finally to Russia – one can only restate Churchill's [1] dictum that 'I cannot forecast the actions of Russia. It is a riddle wrapped in mystery, inside an enigma.' It still retains thus the potential for construction work when politically realisable. Perhaps the newly independent states of the Soviet block offer better prospects with greater security.

Chapter 7, on international strategy, evaluates world markets and the providers of construction services on the world stage. It presents strategic questions as a process where decisions have to be taken about which country to work in. The main issues include politics, economics, culture, currency, etc. What level of project? The issues here include project size, cost and location, language, project administration systems, legal position, etc.

What do we know of our competitive position, including gathering and evaluating data about the likely competition? In short the chapter reviews the factors, issues and criteria to be considered when formulating a strategy for internationalisation.

Part C narrows the discussion to such techniques as may be useful to the strategic planner. Chapter 8 identifies some structured methods for probing the future. In particular the use of expert surveys about the future – Delphi techniques – are described and examples of where they may be used are given. The limitations of the techniques with their tendency to be pessimistic in the short term but optimistic in the long term echo Gramsci's [2] thought that it was necessary to have 'pessimism of the intellect but optimism of the will'. We optimistically see a bright future some distance away and our will draws us forward but we are gloomy in the short term because we can see the problems ahead. The Delphi technique is complemented by an explanation of how scenario planning may be applied to construction. Scenario planning started life at Royal Dutch Shell and was used to present alternative states of world trading. At the time of the Middle East oil crisis in the early 1970s Schwartz [3], the chief strategist for the company, presented to the board several alternative scenarios based upon different stories about how the world may change with the oil price. The board of the company expressed interest in these stories and little happened. Schwartz realised that what he hadn't told the directors is what these stories meant for the company. The lesson is that the scenario has to engage those touched by the possibility of certain outcomes happening. As can be seen scenarios are not planning solutions but are multi-faceted 'devices for ordering one's perceptions about alternative environments in which one's perceptions about alternative environments in which one's decision might be played out' (Van der Heidjen) [4]. As such they are useful instruments for planners in turbulent industries such as construction. These two forecasting techniques are supported by a discussion of portfolio techniques – a mechanism for risk spreading over business areas, markets and project type.

The second chapter in Part C, Chapter 9, is dedicated to marketing and promotional strategies. It catalogues the various approaches to marketing including relational marketing, concepts of service quality and customer care. The chapter argues that the cult of the customer so warmly embraced by companies that are perceived to be successful should be championed by construction companies. It is suggested that customers be classified into key clients (respect orders received for this group), existing non-key clients, infrequent clients and potential clients where a business relationship is not in place.

The marketing strategies need to be supported by activities which promote the company. In part this will engage the firm seeking to establish a 'brand'. This 'branding' of construction companies is a difficult business especially for contractors; it is less so for well known designers. We all know a Frank Geary building or a Norman Foster marque. The design is the imprimatur of a style. The contractor has a less distinctive role, when a building is complete we can see the 'brand' of the designer but we cannot say that it is a Bovis building or a Taylor Woodrow or Laing Construction building. Branding for construction firms has now to come in other ways, through specialisation of the kind of services or project types that are offered. Additionally it is perhaps untimely that just as construction is becoming brand conscious, Naomi Klein's [5] polemic against branded corporations – *'No Logo'* – is becoming influential amongst young people.

The final part of the book seeks to synthesise the theory and the practice. It starts by an evaluation of an analysis of construction markets and how this impacts upon the business environment which surrounds the individual firm. The business environment in which a firm operates will shape the characteristics and nature of the firm, what products and services it brings to the selected markets. In turn the nature of the firm will influence the organisational architecture, how groups or departments are clustered, how the skills of the firm are presented to clients, etc. These symbiotic factors: 'environment' and 'structure' provide the platform for strategy formulation and the chapter discusses how, when and by whom strategy is made. The work draws substantially upon the competitive advantage and strategic plan schools of thought; it recognises that incrementalist, and emergent schools as well as the power paradigm idea have parts to play in the making of strategy.

The adopted strategy will inform marketing plans which in turn will shape workload opportunities. To make the most of these opportunities the resources have to be marshalled throughout the supply chain.

Strategy in consultants is also presented and the business model for services has been under flux for some time. In the same way that we buy our petrol from our grocers and some of our groceries from the petrol filling station, so construction clients have a bewildering array of providers for all kinds of construction services. QSs can provide project management, but so can architects and engineers. Architects can do complete design or be novated after concept design. Contractors provide the comprehensive design to maintenance service. These challenges in a deregulated, neo-liberal construction economy press the larger consultants into thinking about strategic issues in a way never experienced in earlier eras.

Overall the authors' aim has been to look at construction organisations and their strategy in the context of the theories of organisational behaviour, economics and management. These theories have been harnessed to the practical business of getting construction work done.

References

[1] Churchill, W. (1939) Radio Broadcast, BBC Home Service, 1 October.

[2] Gramsci, A. (1986) *Prison Notebooks*. Lawrence and Wishart, London.

[3] Schwartz, P. (1998) *The Art of the Long View*. Wiley, New York.

[4] Van der Heidjen, K. (1996) *Scenarios: The Art of Strategic Conversation*. Wiley, New York.

[5] Klein, N. (2001) *No Logo*. Harper Collins, London.

A Construction and its business environment

2 Construction – an overview of the market

Strategic management of construction organisations has to take place within the context of the fortunes of the construction industry. As many researchers have documented (Hillebrandt & Cannon [1], Shutt [2]) one of the enduring characteristics of the construction industry is its variability of demand. Taking one snapshot of the construction industry, in 1998 the orders received by contractors were 6% higher than in 1997. The DETR recorded that between 1990 and 1997 the industrial workload declined by 23% yet the years 1995–1998 showed a growth in the same market of 18%. Looking at the same years in the commercial market the same volatility can be observed. Between 1990 and 1993 the volume of work in the commercial market fell by 68% and in the period 1995–1998 it had grown by 43% [3].

These examples of variability of demand in many ways shapes the nature of the industry and the firms which operate in it. The first point to note is that the construction industry, unlike most others, is not a single industry but is made up of several different market areas. For purposes of classification it can be divided into four areas:

- Building
- Civil engineering
- Repair and maintenance
- Materials manufacture.

These may be sub-divided into separate market segments such that building is composed of housing, industrial and commercial markets. This fragmentation enables competition for work – even in boom times – to be sharp and the ease of entry into the industry and most sectors of the market stimulates this competition. Moreover the products of the construction industry tend to be fairly homogeneous in that the finished product cannot be visually identified as being the work of a particular builder (although this may not be the case with designers). This means that, in theory, there is a uniform knowledge of the market and what competitors are doing in that market. However, in practice the industry may be seen as a series of overlapping markets which define a particular service, and these markets may be divided by geography, size, type and complexity of work. Thus, put bluntly, there is no competition between a small repair and maintenance builder and a large national contractor; more subtly there are defined market areas for the large national contractors and whilst they may be capable of competing in all sectors of the industry, the strategic choices they make confine them to one or two major market areas. Similarly the small builder may make strategic choices but more often these are based upon location than type of work.

So the construction industry may be broken down into several markets and this provides opportunities for constructors to focus activity or remain flexible to enable them to compete in all sectors. The question thus facing strategic planners is, what are the trends for each sector of the market?

Industrial building

The market for the industrial building sector was erratic and disappointing during the period 1990–1999. This volatility can be seen in Figure 2.1. Despite the recovery of the industrial market from 1993 to 1999 demand is approximately one third of the demand for commercial buildings over the same period. It is considered that several factors seem to be shaping demand for industrial buildings.

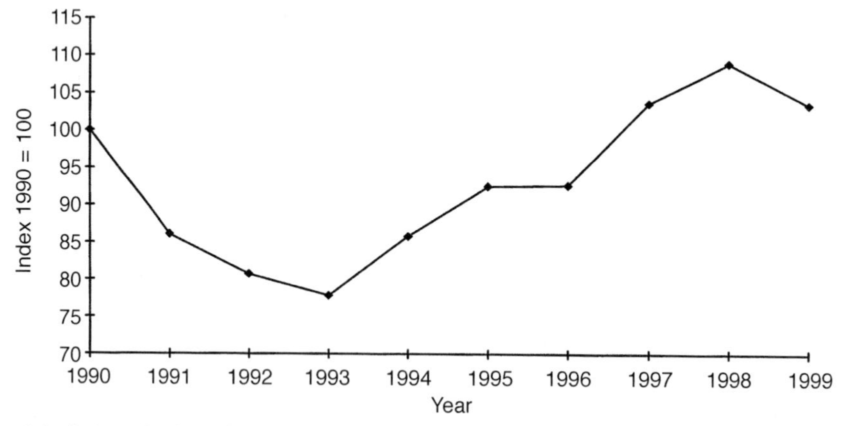

Fig. 2.1 Index of value of industrial work at current prices.

One of the main factors for the decline in demand for industrial buildings during the period 1990–1993 was the erosion of the manufacturing base of British industry and the change from being a country which exports manufactured goods to one which imports such goods. Inevitably the decline in the demand for goods has had a subsequent effect on the manufacturing sector's demand for industrial buildings. However, the increase in demand for industrial buildings between 1993 and 1998 may be attributed to the dramatic influx of foreign industrial companies investing in, and constructing, new production facilities in the United Kingdom. But, as the data from the housing and construction statistics, published by the DETR, shown in Table 2.1 reveal there was a sharp decline in industrial building between 1991 and 1992 although the sector recovered significantly between 1994 and 1999. JFC [4]

Table 2.1 Value of Industrial building by building type

Year	£m	£m	£m
	Factories	Warehouses	All Industrial
1988	2188	751	2763
1989	2741	902	3425
1990	2615	876	3394
1991	2143	561	2622
1992	1790	509	2234
1993	1729	564	2208
1994	1948	626	2489
1995	2368	737	3008
1996	2366	803	3119
1997	2498	1024	3491
1998	2638	1197	3801
1999	2690	1292	3926

Source: Housing and Construction Statistics, The Stationery Office.

forecasts a slowing down of demand for industrial buildings with a marked decline in 1999. This may be attributed to the uncertainty in the business community about the UK's position in respect of membership of the Euro and declining enthusiasm for inward investment amongst hi-tech manufacturers.

Commercial building

There has been a marked change in the demand for commercial buildings in recent years. In the late eighties and early nineties there was a strong demand for commercial buildings from firms operating in the financial services sector although demand in this sector began to wane between 1992 and 1993. However, this sector appears to be recovering, with demand from peripheral financial services sector businesses, such as call centres, creating demand for commercial premises with a particular configuration. The growth in this sector of the economy reflects the changed economic demography: services increasing at the expense of manufacturing; financial capital driving out manufacturing capital with consequent changes in the shape of the labour force; service sector employment taking up the job losses from manufacturing. The commercial market has been dominated by a need for buildings of high quality with space and services provided for high technology office equipment.

External technological developments and the requirements of building users have therefore shaped demand for the products of the construction industry. Demand has been particularly strong in the offices sector of the commercial market and has focused upon new building, as exemplified in the emergence of the commercial 'mega-project'. Such projects have dominated the commercial construction scene during the 1980s, comprising approximately one-third of output. The future of such mega-projects is uncertain, sensitive as they are to shifts in interest rates, political policies, continued demand for office space in the South East and new concerns about their social, cultural and economic sustainability.

An alternative to the mega-project is the renovation and upgrading of offices built in the 1960s to meet new standards of space and comfort as well as provision for new technology. Indeed this may represent a less risky investment, although orders for large-scale projects are often broken down into phases which enable developers to spread the risk element attached to a single large project. Despite the uncertainties of the future, the steady increase in commercial construction during the mid-to-late nineties can be seen in Figure 2.2. The property boom of the eighties, which saw a peak of £11 310m in 1990, gave way to a steady decline in the early nineties to a low of approximately £5131m in 1993. Growth in this sector has been pronounced in the period 1995–1998 However, it would seem that the polarised extremes of the market in the eighties have given way to a more balanced demand with the forecast showing a very gradual growth.

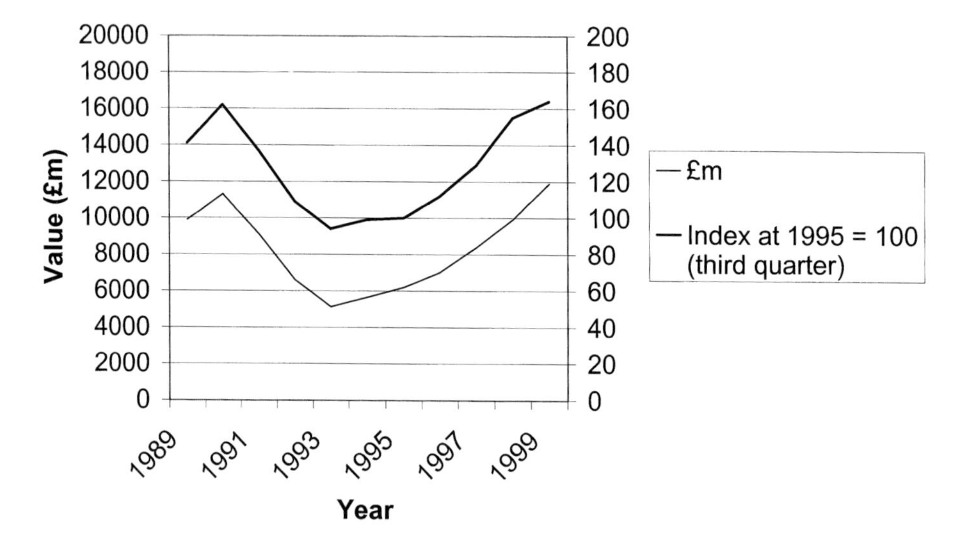

Fig. 2.2

The housing market

In 1999 the market for the provision of new housing (both public and private) represented about 28% of total construction output by value. The proportion of housing value has fallen steadily since the early 1970s when it represented some 30% of the value of all new work. This is, of course, a reflection of the changing pattern of demand, with the boom in commercial building depressing the proportion of housing. However, the trend has been for the total number of housing starts to remain relatively constant over the period 1990–1998. This trend masks the real changes in the housing market. Total public sector housing starts in 1998 (30 000) are virtually the same as those in 1986 but by 1998 the number of local authority houses started had collapsed to 482.

Clearly there have been seismic changes in the housing market. Changes wrought by political doctrine asserting private provisions were to be preferred over public support. Moreover with the enactment of the Housing Act 1990 the role of local authorities in the provision of housing changed dramatically: this act has supplanted the role of the local authority as a housing provider with that of private landlords and housing associations. Table 2.2 shows that local authority housing starts have declined from 9000 in 1990 to approximately 500 in 1998 whereas housing association housing starts have increased from 8 000 in 1990 to a peak of 44 000 in 1993 but have declined to approximately 24 000 in 1998. In contrast, private sector housing starts have remained relatively buoyant with a gradual increase from 137 000 in 1990 to 150 000 in 1999.

Table 2.2 Housing starts and completions in Great Britain (000's)

	1990	1991	1992	1993	1994	1995	1996	1997	1998	1999
Public sector										
Starts										
Housing associations	18	22	34	42	41	32	29	26	24	
Local authorities	9	4	3	2	1	1	1	1	0.5	
Total	27	26	37	44	42	33	30	28	29	35
Completions										
Housing associations	17	20	26	35	36	38	33		24	
Local authorities	17	10	5	3	2	2	1			
Total	34	30	31	38	38	40	34	27	27	29
Private sector										
Starts	137	137	120	141	158	134	145	160	155	150
Completions	160	154	141	141	147	149	147	150	160	150

Source: JFC Construction Forecast [4].

The forecasts shown in Table 2.2 show that total public sector housing starts are expected to increase marginally to 35 000 per year while private sector starts are expected to increase marginally then gradually decline slightly. Figures 2.3 and 2.4 respectively chart the resurgence of public sector housing starts, brought about by the growth of housing associations as social housing providers, and the cyclic boom/bust nature of the private sector housing market in the post-eighties period. Obviously Government fiscal policies in respect of public sector spending and the restrictions on local authority housing projects have shaped the type of demand. But this would suggest that it is merely economic issues that shape the character of demand; political choices are made in respect of housing and these choices are ideological in nature.

Personal values have been changed by the political and social process so rather than the late-nineteenth-century view that 'we're all socialists now' the

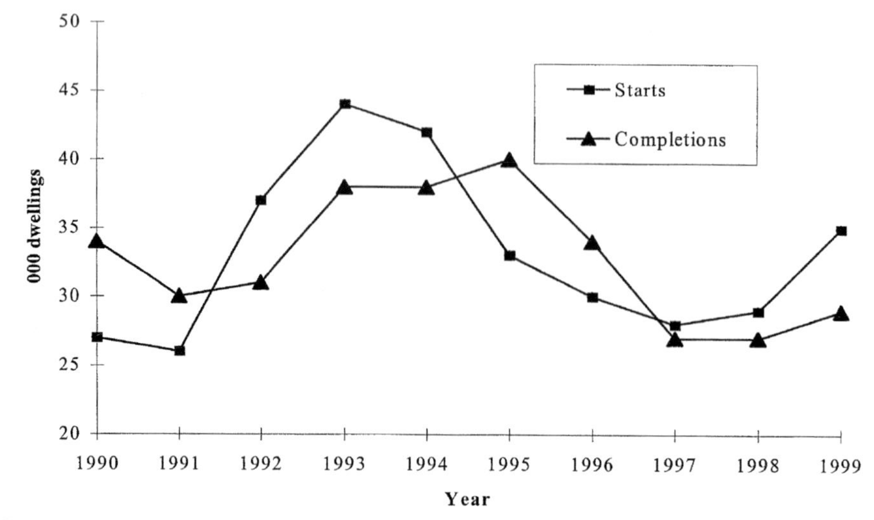

Fig. 2.3 Public sector housing.

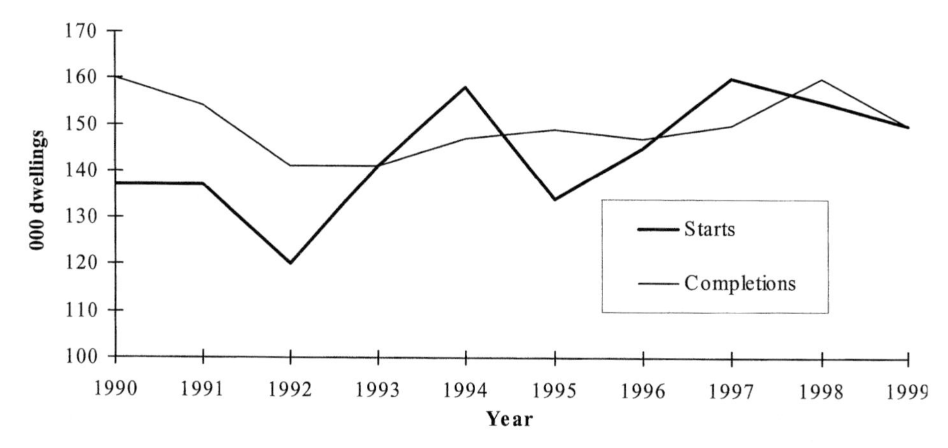

Fig. 2.4 Private sector housing.

dominant set of cultural values of the 1980s and early 1990s has been for individualism to drive out the traditional values of collective societal responsibilities. The election of a Labour government in 1997 did not change the general preference for private over public provisions and the government, of whichever party, have been committed to private finance initiatives (PFIs) for delivering public services. In the context of housing the broader impact of PFIs is discussed in the next chapter. This changed set of values can be seen as a paradigm shift which has forged preferences in models of housing provision. All these factors have a bearing on the changed pattern of the housing market and strategic planners in construction need to be aware of signals which predicate such changes.

Repairs and maintenance

During the period 1991–1998 the overall expenditure on repair and maintenance output, when allowing for inflation, remained virtually constant (£21 900 million in 1991 and £24 000 million in 1998). This relatively constant level of total expenditure is not wholly accurate. Public housing repair and maintenance increased by about 8% over the period 1991–1998 as a result of the release of frozen local authorities' capital receipts. In the private housing sector, repair and maintenance decreased by about 47% over the same period and indications are that this trend may continue into the next few years as a result of the increase in house prices. Demand in private R&M work in the late 1990s was fuelled by de-mutualisation windfalls in the period 1996–1997. Public non-residential repair and maintenance suffered from a 19% decline over the period 1991–1996 and looks set to continue. This is largely due to the severe financial constraints faced by local authorities. Private sector non-residential repair and maintenance stayed relatively constant over the period 1991–1995 but exhibited a significant rise of 26% between 1995 and 1998. This was largely due to rising corporate profitability and the desire to minimise tax liabilities.

Making sense of the trends

The strategic planner must assess the significance of these trends within the construction industry. It may be useful to consolidate the data to identify the major issues which are shaping demand and then explore the impact of these trends on construction firms. The first point to make is that there are two indicators that the planner can use – the value of new orders and the value of construction output. Orders are based upon contractors' returns whilst output is an aggregate of all construction work and would include estimates of the, now significant, amount of work undertaken in the frequently unrecorded repairs and maintenance sector. Therefore output is the indicator more frequently used in any data analysis.

One of the starkest trends in output is the decline in public sector work which began in the 1970s. This decline has, however, been compensated for by a growth in private sector work. Consequently changes in total output may be relatively small because changes in different sectors of the construction market have cancelled each other out. Such patterns have implications for planners in firms that specialise in particular sectors of the market. Specialised firms may suffer from a temporary shortage of work in their area and there may be barriers preventing them from transferring to more buoyant sectors of the market. These may be a lack of technical ability or the geographical distance of a project from the usual area of work. The geographical distance may deprive a firm of important local contacts with clients, designers and sub-contractors.

The geographical distribution of output is a factor worthy of consideration. Bluntly stated, the statistics show that over the period 1986–1999 the regional balance of new orders in the UK has remained relatively constant with the South East (including London) accounting for the greatest share of new orders. Thus the market for construction is dominated by the South. Figure 2.5 shows the balance of work in the two regions from 1986 to 1998. It can be seen that the South has the lion's share of value of new orders obtained over the whole period. In 1986 the South East obtained over four times the value of new orders obtained in the North West although this had dropped to just over three times as many new orders by 1998. From Figure 2.5 it can be seen that new orders obtained in the two regions show broadly similar patterns. The boom of the late eighties, which was dramatic in the South East, had a very minimal effect in the North West. Similarly, the significant relative decline experienced in the early nineties in the South East only resulted in a marginal decline in the North West. A strong message from this is that, while the South East takes the lion's share of the balance of new orders, the construction market in the South East is very sensitive to economic cycles. This is further emphasised in the distribution of National Lottery money for millennium projects which had a construction content. Evidence would suggest that Lottery money has been focused on major projects in London and the South East.

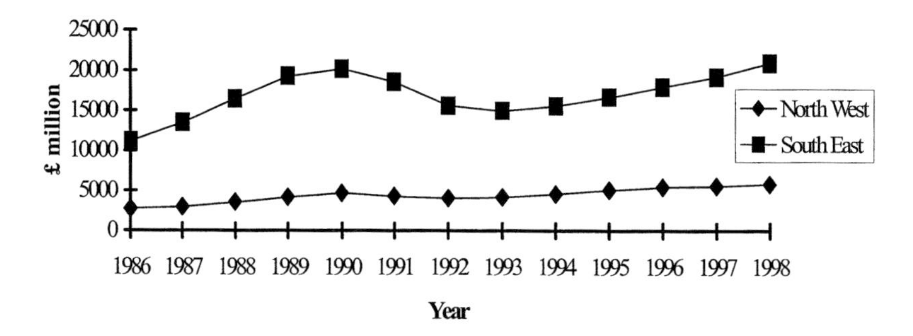

Fig. 2.5 Value of work by region (1986–1998).

What are the implications of these trends? The decline in the public sector workload meant that firms who specialised in public sector works in the 1960s had to drastically restructure in the 1970s to be in shape for the pattern of work in the 1980s. Those benefitting from the communal developments of the 1980s had to restructure and reinvent themselves to accommodate PFI work in the 1990s and 2000s. Not only has this change had implications for the structure of firms, it has also changed the balance of staff within firms. Both public and private sector projects are now characterised by the predominance of projects which last for 12 months or less. This has, historically, been the picture for private sector housing projects but not public sector projects. In the early to mid eighties it was found that public sector projects frequently had durations of greater than 24 months: this is now not the case. It would appear that the severe restriction of public sector finances and the subsequent diminution of local authorities in procuring construction projects have meant that public sector projects tend to be on a much smaller scale and hence have a much smaller duration. It has been shown earlier that repair and maintenance work is a steady workload for the public sector which now appears to be refurbishing and maintaining its property stock rather than commissioning new stock. This work tends to be relatively short term, i.e. having a duration of less than 6 months. Figures 2.6a–2.6e show the value of new orders obtained in 1998 categorised by project duration.

It can be seen that public sector housing projects mostly have a duration of less than 12 months (over 90%) whereas 75% of private sector housing projects have a duration of less than 6 months. Other public sector projects mostly have a duration of less than 6 months: this confirms the view expressed above that

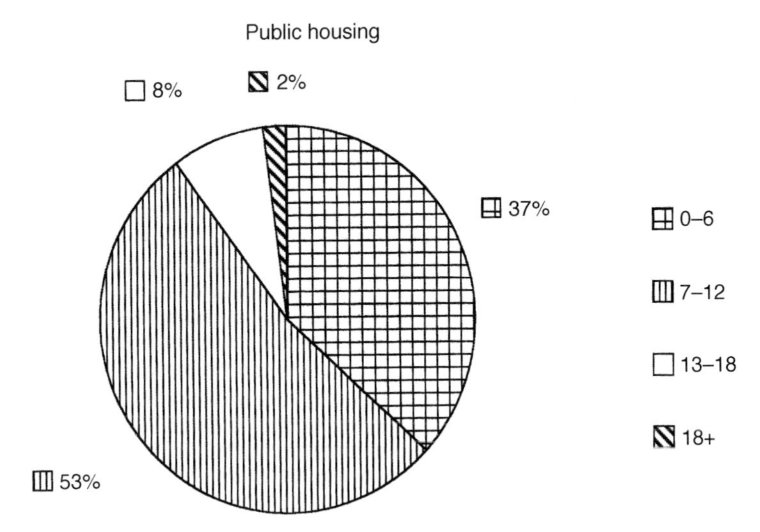

Fig. 2.6a 1998 number of new orders by size of projects (months).

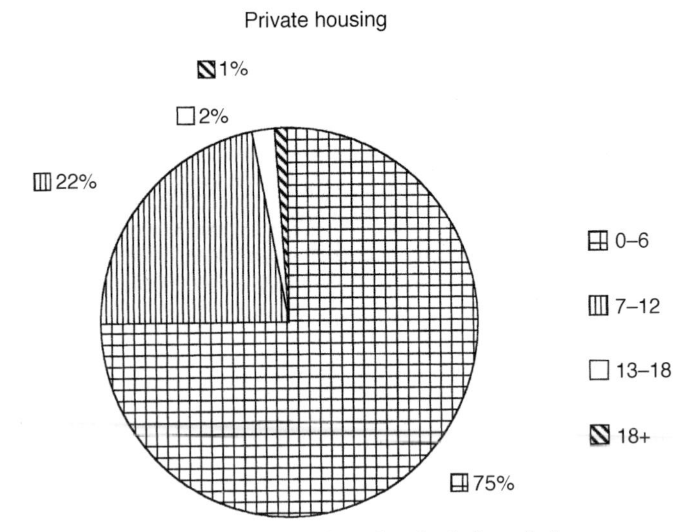

Fig. 2.6b 1998 number of new orders by size of projects (months).

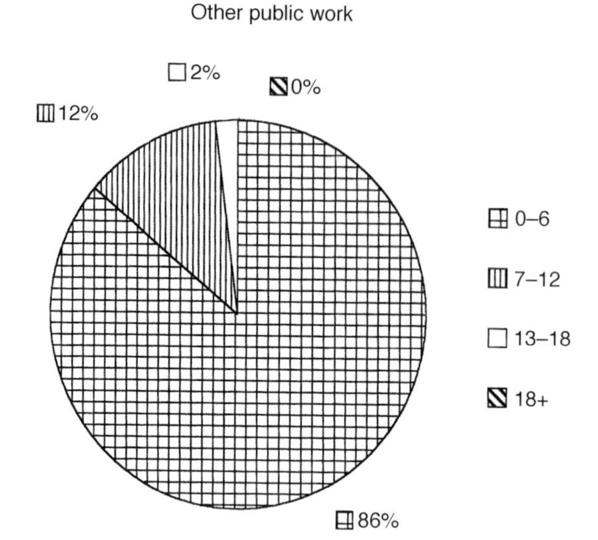

Fig. 2.6c 1998 number of new orders by size of projects (months).

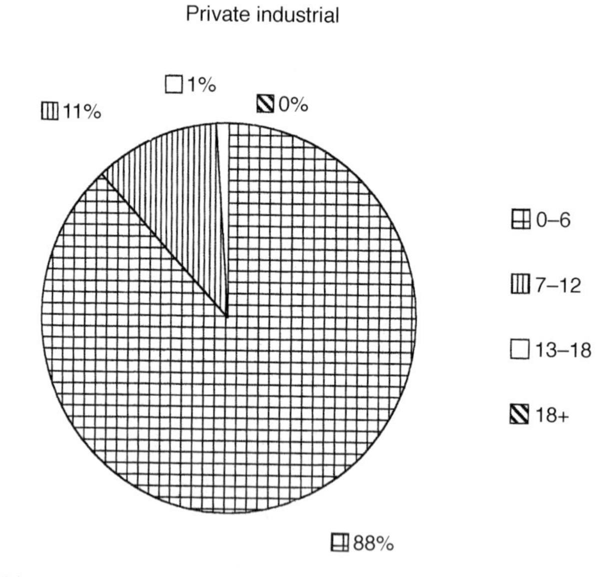

Fig. 2.6d 1998 number of new orders by size of projects (months).

most public sector projects are now of a repair and maintenance nature. This picture is repeated for the private industrial and private commercial sectors which, respectively, have 88% and 91% of projects having a duration of less than 6 months. These figures are in marked contrast to the position in the mid-eighties where approximately 86% of private industrial projects and approxi-

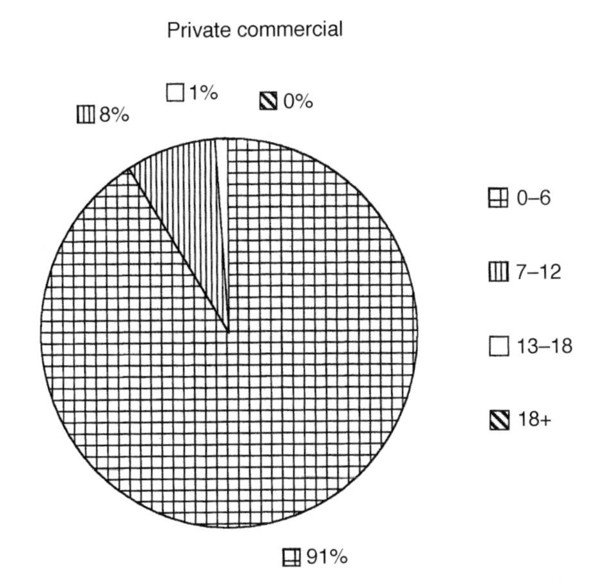

Fig. 2.6e 1998 number of new orders by size of projects (months).

mately 89% of private commercial projects had durations of less than 24 months. The main reasons for this paradigm shift are the decline in major industrial projects coupled with an increase in smaller industrial facilities: with a corresponding reduction in project duration and a major reduction in private commercial mega-projects. Figures 2.7a–2.7e show the value of new orders in 1998 categorised by value of individual projects.

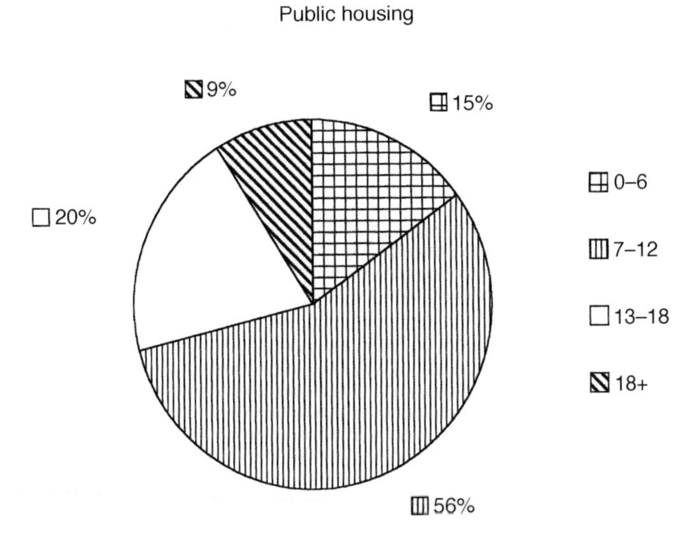

Fig. 2.7a Value of new orders – 1998 displayed by the size of projects (data expressed as percentage of total value).

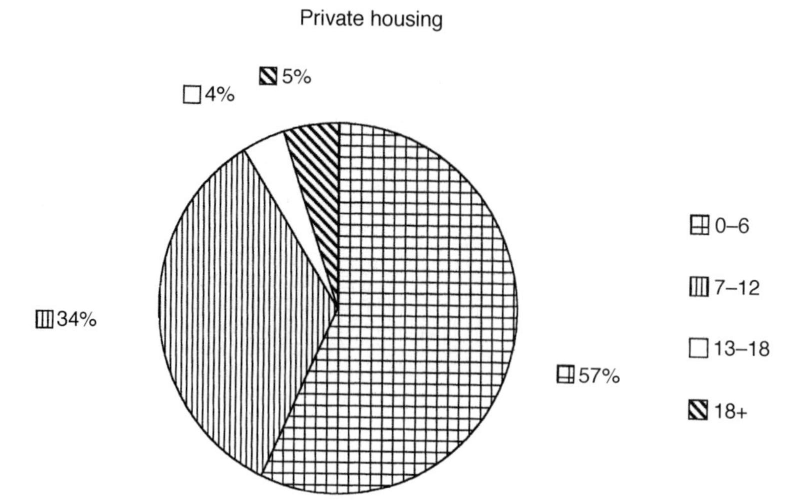

Fig. 2.7b Value of new orders – 1998 displayed by size of projects (data expressed as percentage of total value).

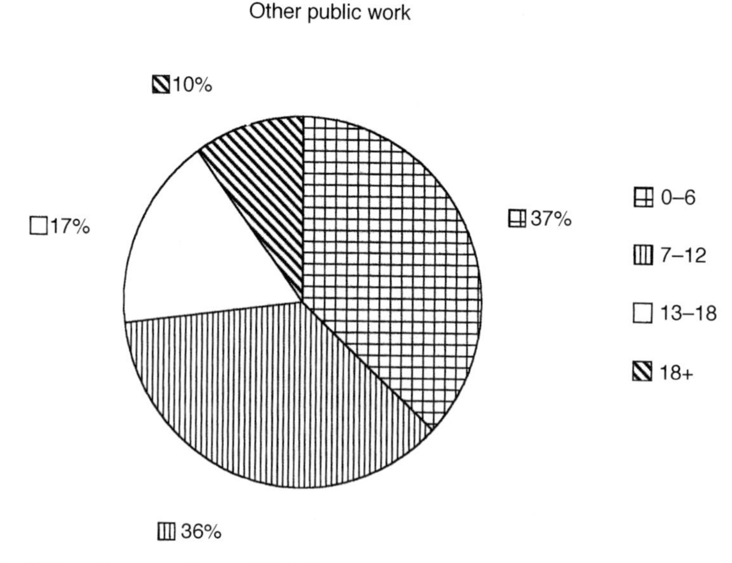

Fig. 2.7c Value of new orders – 1998 displayed by size of projects (data expressed as percentage of total value).

Inspection of the DETR data shows that over half of public sector housing projects ordered in 1998 had a project value of less than £2 million whereas private sector housing projects of less than £2 million accounted for almost 91% of orders placed in 1998. Private industrial projects were almost evenly split between those less than £2 million (55.9%) and those greater than £2 million

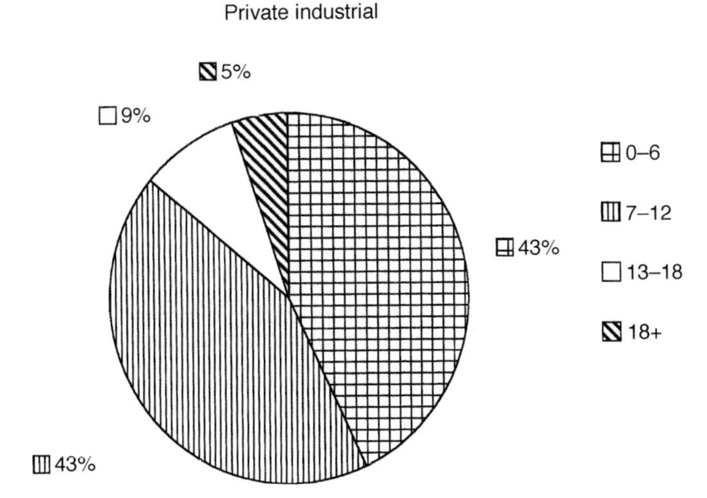

Fig. 2.7d Value of new orders – 1998 displayed by size of projects (data expressed as percentage of total value).

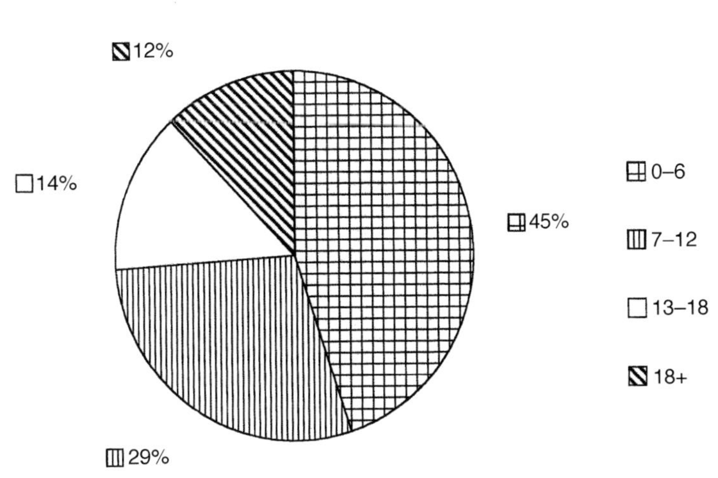

Fig. 2.7e Value of new orders – 1998 displayed by size of projects (data expressed as percent of total value).

(44.1%). Private sector commercial projects greater than £2 million accounted for almost 55% of orders placed in 1998 with 38.1% being between £0.1 million and £2 million. Ball (5) notes that 'only large office blocks and energy related ventures match the value and time scale of public sector work, and they are a quite small proportion of private sector work'. This is as valid now as when the observations was first made.

This trend towards shorter duration projects has meant that the ratio of

Administrative, Professional, Technical and Clerical (APTC) staff to operatives had to increase and this has had implications for the personnel policy of firms. The change in the proportion of APTC staff as a percentage of the operative workforce in direct employment is shown in Figure 2.8. The sharp decline in the ration of APTC staff to directly-employed operatives in the period 1997–1998 has been caused by changes in the employment status of operatives. Around 1996 the Inland Revenue were much more vigilant about the status of the genuinely self-employed. Consequently many former 'lump' workers with long associations with particular construction firms were put back on the employment books of construction firms. Hence the numbers of APTC staff remain reasonably static whilst the numbers of directly-employed operatives increase at the expense of the numbers of self-employed. This factor explains the break in the trend line.

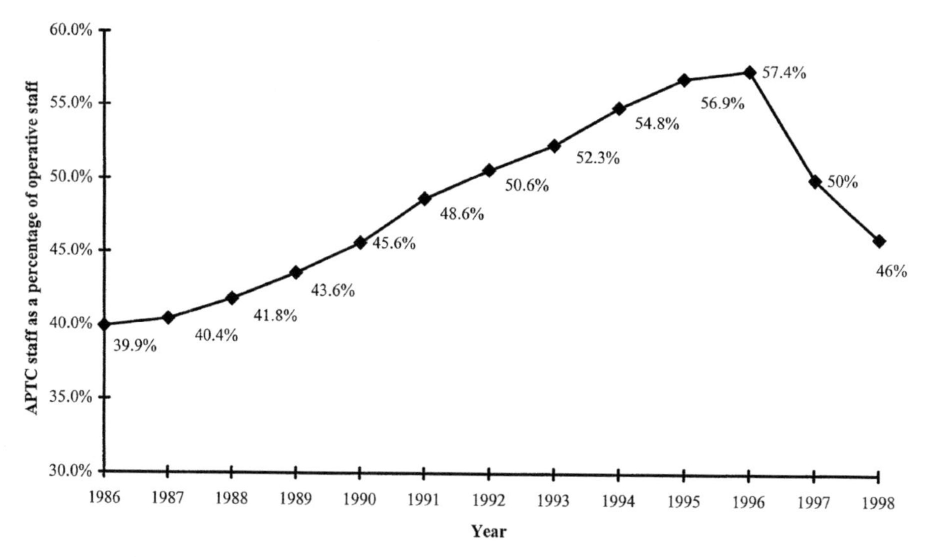

Fig. 2.8 Ratio of Administrative, Professional, Technical and Clerical (APTC) staff to operatives.

A further feature of the trend towards private sector work is that much of the commercial and industrial work, as well as the housing work, is speculative. This has required land banks to be provided and has resulted in capital being tied up. Thus the maxim of highly flexible capital for building firms has had to be re-appraised. This capital lock up is amplified by equity stakes on PFI projects.

It is also noticeable that 'demand in the construction industry is created in a different way from in many manufacturing industries'. The nature of competition in construction is determined substantially by the actions of the client and consultant. Demand is part of the environment to which senior managers must respond and the nature of demand in construction is characterised by the following [6]:

(1) The exercise of monopsonist power by clients for construction to contract in the precontract stages. Power shifts to the contractor who is responsible for the effective and efficient on-site production process. Once the contract has been signed the client and the client's advisers have little or no direct control until the contract completion date is reached, although options to reintroduce control exist but will probably result in litigation unless the contractor has gone bankrupt.
(2) A shift to private sector demand, which is more variable in the short to medium term and involves contractors in speculative construction to create demand, more land assembly than in the past or contracting through tendering activity to maintain or enhance turnover.
(3) Industry sectoral workloads that may vary and counterbalance each other to provide an aggregate industry workload.
(4) A consistent and upward trend in the workload for repair and maintenance (R&M).

These changing conditions predicate considerable problems with forecasts of demand in construction in the short to medium term, and Hillebrandt [6] argues in favour of understanding need, which she defines as the difference between the total requirement for built facilities and infrastructure and the existing provision, and those factors that contribute to it, namely,

- the populations of users of built facilities and any demographic changes to those populations;
- the rate of usage of built facilities and the change in the rate of usage;
- standards for facilities and any changes in standards;
- replacement of stock due to ageing or technical factors;
- increases in or replacement of stock due to technological change or changes in standard.

Changes in short to medium term demand, which are difficult to predict, fall within the operational domain of managers' thinking, whereas strategic decision-making, normally taken to involve taking a long-term view, with an understanding of the underlying causes of demand conditions, namely, the determinants of market need. Stokes [7] has argued that understanding the trends in the national and international construction industries should be the primary focus of strategists in construction. Also, trends in national fixed capital formation, in which construction forms a major contributing part, provide the backcloth for assessing industry trends and possibilities. The difficulty for the strategic planner in construction is synchronising strategic moves of the company in the short to medium term (2–5 years) with the long trend in output and new orders. In the long term output is relatively stable but in the short to medium term it is quite volatile. Ball [5] proposes that variations are not felt at the aggregate level of industry workload but in shifts within the composition in the sub-market.

As a final point, construction is often reported as being a fragmented industry. A fragmented industry is one in which no company has a significant market share and is able to influence considerable outcomes within the industry. A large number of small and medium-sized companies and a small number of large companies usually populate a fragmented industrial structure. There is also a high incidence of privately owned companies. These defining features are common to construction. Porter [8] adds that a fragmented industry is populated by many competitors who are in a weak bargaining position with respect to both buyer and supplier groupings, and profitability is marginal. Again, this is a defining feature of construction. Male [9] used Porter's sixteen characteristics of fragmentation and concluded that in the region of thirteen were active in construction. The conclusion from the analysis is clear, in general terms the contracting industry is fragmented for a divers range of reasons but the underlying causes are rooted in the pre-demanded, localised project-based production characteristics of construction.

Contracting is also a geographically dispersed project-based industry with markets that operate from the local to the international level. As project size, complexity, technology and international location enter the frame of reference, there are fewer and fewer companies able to undertake particular types of project and fragmentation tends to decrease as the industry becomes segmented within these overlapping project-based market structures. The concept of 'contestable markets' used by Flanagan & Norman [10] in analysing tendering strategy and behaviour is of direct relevance in explaining firm behaviour in the contracting industry's project-based market structures. A contestable market is one where oligopolistic competition operates and where the danger of a potential entrant constrains companies' behaviour such that their pricing policy is affected. In the contracting industry, where the intangible institutionalised distribution channels exist for contractors' services contestable markets operate where projects are large in size, are complex, either organisationally or technologically, or both, and especially where an international dimension to tendering exists. Construction is also an hierarchical industry designated by size of firm, where the many small companies are tending to act as sub-contractors to the large companies (Ball [5]). It is at the small firm end that fragmentation is most prominent.

Therefore, the construction industry is first, comprised of geographically dispersed and overlapping project-based market structures and second, is hierarchically structured in terms of company size. Fragmentation is high at the smaller end of the industrial structure, for example, in repair and maintenance work whilst for new-build work fragmentation decreases according to project characteristics, including those with an international dimension. Entry and exit barriers to the industry are many but exist often in a subtle form. This has established the setting within which managers in construction must decide on the future direction of the firm. The next section draws the various themes together.

In conclusion, it is frequently difficult for firms to shape their organisation to meet the changing composition of the construction markets. Large firms with specialist divisions and regional offices can manipulate resources to compete in a fashionable sector of the market, be it fashionable by type of project or services offered. Firms without a specialist division or a regional structure, however, can find it difficult to break into new market areas. The traditional medium-sized building firm, based in a depressed region, has been worst hit by changes in demand. Strategic planning could equally have helped these firms to survive during periods of market turbulence.

References

[1] Hillebrandt, P. & Cannon, J. (1994) *The Modern Construction Firm*. Macmillan, London.

[2] Shutt, P. (1988) *The Economics of Construction*. Longman, Harlow.

[3] Department of Environment Transport & Regions (1999) *Housing & Construction Statistics* 1988–1998. The Stationery Office, London.

[4] Construction Forecasting and Research (1999) *Construction Forecasts 1999, 2000, 2001*. C.F.R., London. **5**, Issue 3.

[5] Ball, M. (1988) *Rebuilding Construction: Economic Change in the British Construction Industry*. Routledge, London.

[6] Hillebrandt, P.M. (1984) *Analysis of the British Construction Industry*. Macmillan, London.

[7] Stokes, F.H. (1982) *Practical problems in corporate planning: John Laing in Corporate Strategy and Planning*. B. Taylor, J. Sparkes (eds). Heinemann, Oxford.

[8] Porter, M.E. (1980) *Competitive Strategy: Techniques for Analysing Industries and Competitors*. Free Press, New York.

[9] Male, S.P. (1991) Strategic Management in Construction: conceptual Foundations. In: *Competitive Advantage in Construction*, S.P. Male & R.K. Stocks (eds). Butterworth-Heinemann, Oxford.

[10] Flanagan, R. & Norman, G. (1993) *Risk Management in Construction*. Blackwell Science, Oxford.

3 The strategic role of the actors in the construction process

The previous chapters considered the broader dimensions of the business environment in which the construction industry has to swim. The earlier discussions have analysed the sub markets in terms of the demands made upon construction firms along with an analysis of how firms strategically respond to these demands. This chapter provides an anatomy of the players who are making these strategic responses.

The external environment is anything considered to be outside the boundaries of the company and the firm's environment can be broken down into first, the general environment and second, the task environment [1]. The general environment is anything that can potentially or indirectly affect the company and would include the economy, demographic variables, society's values and attitudes, and technological change at the societal level. Other variables would include the legal and cultural framework of a country [2]. The task environment comprises those variables, and only those, that would have a direct and immediate impact on the company's activities, it is the primary source of opportunities and threats for a company. Due to its direct impact on the firm it is likely to be better understood than the general environment [3].

The evolution of the construction industry in the post World War II era

The industry environment of the 1950s and 1960s for contractors was characterised by stability reflected in high growth rates in construction output. Periods of recession only slowed the rate of increase. This rapid growth stemmed from the level of investment in infrastructure projects and house building. During this period of stability firms used their experience to identify those markets to which they were most suited. The strategic management process focused on systems and procedures that would facilitate either control and/or integration within organisational structures and provide competitive advantage by focusing on internal efficiency since markets would not change dramatically. Market specialisation occurred, not as any premeditated strategy, but rather through the use of managers' experience to identify types of projects which were successful for firms. The strategy formulation process was simple with little need for reference to the external environment. Lansley [4] asserts that the implicit focus of contractors during this period was the utilisation of a focus strategy based on cost leadership. The 1950s and 1960s were therefore characterised by operational change.

Ball [5] asserts that the years 1968/69 formed a watershed for the construction industry. New orders in the sectors that had underpinned the earlier boom dropped dramatically as a result of the economic crisis of 1967, the devaluation of the pound and subsequent government policy. In the depressed conditions of the late 1960s contractors were forced to make rapid moves into new markets in order to avoid a sharp downturn in profits and as a consequence there was a merger boom in the late 1960s in response to falling workloads.

The 1970s were characterised by strategic change. The most important change was the sharp and persistent fall in demand. There was a sustained fall in total demand for construction of 30% in some markets and 90% in others. These changes were in complete contrast to those that had taken place in the preceding two decades. The dramatic changes in the environment of the 1970s meant that specialisation had to give way to a strategy of breadth in order that firms could cope with the uncertainties and troughs in demand. These changes required a strategic orientation and a need for creative insight by senior managers. The problem facing contractors was the need to move away from a focus strategy to one of competing on cost within the context of a broad market base. The requirement for managerial decision-making and thus for company strategy was flexibility. A number of important strategic decisions were taken within firms during this period. Decisions taken in the 1960s to vertically integrate were reversed and companies divested themselves of businesses considered outside their 'core' activity. Portfolio management became important. The civil engineering sector, which had grown rapidly in the 1950s and 1960s, especially with motorway construction, faced a slowdown in growth in the 1970s and virtually all firms in this sector attempted to diversify.

A strategy based on steady internal expansion, of relevance in the 1950s and 1960s, was no longer an option in the 1970s. Acquisition was the only way of increasing market share. A merger boom occurred in the early 1970s as a result of a rise in new orders for certain types of work. However, by 1974 demand in every sector of the industry had plummeted. Takeover activity also fell dramatically and remained low until 1978. The crisis did not have a uniform impact for some contractors continued to make record profits, but not on the magnitude of orders won. Many firms diversified overseas or adopted a 'wait and see' policy.

Additionally during the 1970s, takeover activity changed in terms of size of firm involved. In the early 1970s takeover activity was dominated by small firms acquiring others and remained that way until the slump of 1974. The medium-sized firms dominated acquisition after 1974. This reflected, in part, the increased size of firm which resulted from the earlier small firm activity. From 1976 onwards large firms also increased their merger activity. By the 1980s some firms were so successful in diversifying that contracting represented only a small percentage of their overall business. They had become construction corporations, some having structural configurations that more closely represented construction-oriented conglomerates.

The issues confronted by managers in the 1980s were not unfamiliar and had their roots in events occurring between the 1950s and 1970s. The environmental circumstances in the period 1980–2000 for contractors included:

- Competitive changes in the environment largely due to long-term shifts in the structure of the industry and in client attitudes towards the industry;
- The increasing complexity of large building projects;
- The move away from traditional forms of building contract;
- Rapidly declining workloads with demand becoming unpredictable although there were short-lived recoveries in market by project type and location;
- A lack of skilled labour;
- A fall in overseas opportunities;
- New forms of project financing coupled with lack of public finance;
- Government pressures through Latham and Egan reports for performance improvements;
- Partnering and less adversarial cultures developed.

Lansley's [4] contention, and one that presumably held for the late 1990s, is that contractors were not well-versed in managing in such an environment. Therefore, in situations where demand becomes less predictable a policy of sub-contracting has considerable merit. The implications of these trends for the strategies of contracting firms in particular are clear. Specialisation in one sector could be fraught with considerable difficulties since work shortages may be inevitable. The counter to this is to spread risks and operate in a number of sectors. However, the technologies required in other sectors may be different. This favours the large contractor who has the resources to cover such a range of work. Straddling different industry sectors typifies the strategic behaviour of the largest construction firms in the industry. These firms emphasise particular sectors in accordance with current management strategies. In addition, if, as Ball [5] contends, contractors act in a merchant-producer role, this, combined with their market power, allows them during periods of slump to pass costs onto others. Workload flexibility puts contractors, therefore, in a strong position over sub-contractors, materials suppliers and building workers and provides them with the ability to maximise sectoral market opportunities.

The common theme running through this period change in the postwar period is the turbulence of the environment in which construction has had to operate. An analysis of this environment is therefore an important part of the analysis.

Lansley *et al.* [6], whilst researching the survival strategies of contractors, provided a useful analytical framework for thinking about the external environment in construction. He divided the external environment into three broad categories. First, the *common industry/national environment* comprising the economic and social background of firms, the industry's existing and potential clients, suppliers, labour and respective trade unions, trade associations, central and local government departments. This is common to all firms. The *competitive environment* is more 'localised' to the firm and describes the environment where the company is in competition with other firms. Key factors here include the structure of demand, competitors, availability of materials, labour, sub-contractors and suppliers. Lansley *et al.* noted that as company size increases, the boundary between the competitive and industry environments becomes blurred. The *operational environment* is unique to each company. It deals with the position of the firm in the competitive environment and has its origins in the strategic choices made by the company, the business activities which it undertakes, the geographic areas within which it chooses to compete and the suppliers with which it deals. The box below draws the environment characteristics together into an analytical framework.

Environmental Types		Characteristics	Comments
General environment	Common industry/ national environment	The economic and social background of firms, the industry's existing and potential clients, suppliers, labour and respective trade unions, trade associations, central and local government departments, professional institutions, industry Task Forces and initiatives	Common to all firms in the industry and impacts firms both directly and indirectly. Impacted by demographics, technological and societal changes
	Competitive environment	Structure of demand, procurement forms used by clients, competitors, and availability of materials, labour, sub-contractors and suppliers	Localised to the firm. The competitive environment is determined by Porter's five industry forces introduced in the section in Chapter 10 dealing with industries and markets
Task environment	Operational environment	Positioning of the firm in its competitive environment. Determined by strategic choice and business scope, and the suppliers with which it deals	Unique to each firm

Repair and Maintenance (R&M) has had a consistent upward trend and work opportunities in this sector are therefore considerably better. Methods of working and contract management requirements are different from those for new building projects. The scale of work operation for R&M favours the small contractor.

The net effect of client pressures during this period has been a demand for a greater diversity of services and ways of interfacing with the industry, especially for contractors. One of the major issues for clients in the 1980s was that of quality in construction, by the 1990s the focus had moved to benchmarking partnering and performance indicators. This concern for quality has affected not only contractors but also the professions. For contractors, the main issue highlighted in this chapter has been the use of diversification to counterbalance falls in sectoral demand by switching into different sectors to take advantage of their upswings. The financial flexibility of contractors allows them to diversify through either internal expansion or acquisition.

The contractors

A simple way to outline the nature of the industry is to define it by output and by the number of firms producing construction work. Indeed, the 170 000 heterogeneous and fragmented firms undertaking some £60 billion of work each year are one way of defining the industry. The workload undertaken by these firms typically includes general construction and demolition work, construction and repair of buildings, civil engineering works and installation of fixtures and fittings. This work is undertaken by a large number of small firms with a small number of large firms competing for the largest projects. This suggests that the construction industry comprises firms who differ in terms of size and scope, and even within firms there is often a great diversity of activity with different parts of the firm tackling specific sub-markets. The very size of the construction industry makes it an important economic entity that employs around two million people directly or indirectly in construction-related industries, constituting some 7% of the total UK workforce. Equally the construction industry provides approximately 7% of the Gross National Product (GNP).

Construction is therefore essentially a large industry of small firms. It is staffed by operatives who are predominantly young, male and casually employed, with historically a strong craft tradition although current construction processes are replacing this tradition with a return to prefabricated components being installed by workers offering new sets of skills at the site, the craft process being transferred to off-site fabrication centres. These changes have wrought massive realignments in the construction production process, the old tradition of design as a separate entity from production is waning in the face of integration of the process by design and build arrangements and prime contracting where 'turnkey' approaches are evident.

This separation has important ramifications for the classification of the construction industry: design, quite properly, can be seen as a service industry but is construction better defined as 'manufacturing'? Given the nature and diversity of activities carried out in the construction industry, there is a natural tendency for the industry to be viewed as manufacturing in nature rather than a service industry. Newcombe [7] submitted that 'it is a misnomer to classify the construction industry as a service industry along with banking, insurance and retailing'. The definition of construction is important as the way in which one looks at the industry defines its markets, and consequently, the strategic processes which are used to govern and direct the construction organisation. Newcombe's analysis suggests that the principal functions performed in the manufacturing industries can be mirrored in the construction industry although he uses different titles for various functions. He claims that the functions carried out in construction are comparable to those of manufacture as illustrated in Table 3.1. This view is not orthodox.

Table 3.1 Principal functions performed in the manufacturing and construction industries

Manufacturing	Construction	Principle of Function
Marketing	Estimating	Identification/creation of markets, and selling of end 'products'
Production	Construction	Organisation, movement and assembling of various materials, components, etc.
Purchasing	Buying	Acquisition, bulk or otherwise, of production materials and components for a project or in lieu of a project

Source: Newcombe (1976) [7]

Fleming [8] relates the classification of the industry to the production processes used in construction and manufacture:

> 'the relative and absolute cost advantages which often favour large scale operations and large established firms in manufacturing are of little importance ... as factors encouraging the growth of greater industrial concentration in construction... The site based nature of construction where each site is necessarily a temporary place of work, and the individuality of most projects ensures that the conditions necessary for the existence of many technical scale economies, mainly centralization of production of standard products using specialized production techniques do not apply.'

This view is supported by Hillebrandt [9] who considers construction to be a service industry. This conclusion is drawn from the evidence of what builders actually do. She notes that 'construction may be regarded as one industry whose total product is durable buildings and works'. It is 'the contracting part of the industry which undertakes to organize, move and assemble various materials and component parts so that they form a composite whole of a building or other work. The product which the contracting industry is providing is basically the service of moving earth and material, of assembling and managing the whole business'. This observation recognises the changes that have been taking place in the construction industry over the last twenty years, resulting in a strong differentiation between contractors who provide management services and contractors who undertake to build the physical product.

The separation of the industry into these two distinct areas has been one industrial response to the relatively high levels of risk which are perceived to exist within the market. This risk is being passed down the line to those who actually do the construction work. The separation of the management and the doing is reflected in the technical processes involved in each aspect of construction work and, therefore, industrial classification may be based upon differences in the technical processes undertaken. Large firms providing management contracting and project management services may be regarded as part of the service industry, whereas those providing resources which are used to construct the building might be better described as manufacturing-style organisations.

A further complication exists when large organisations are highly diverse in their activities. Those firms designated as building and civil engineering contractors offer quite distinct services which operate in distinct markets and their staff may not be generally interchangeable. Large construction organisations may be involved in the building of houses for sale, which again forms a different market and requires different sets of resources and management skills. Many contractors have moved into property investment as a mechanism for forward integration. They have also sought to move backwards into product manufacture in order to stabilise the environment in which they work. Contractors are therefore seeking to spread risks by strategies of related diversification into connected markets. Catherwood [10] distils this approach rather succinctly when observing that 'a successful general contractor may fail miserably in speculative house building ... or in civil engineering, conversely a civil engineering contractor may undertake a building contract at his peril'.

In the light of the above observations the construction industry may be defined in terms of several distinct construction markets some of which provide a service, others which reveal characteristics of a manufacturing organisation. A typical classification divides these into five business arenas, principally: civil engineering, building, property development, estate development (housing) and construction product manufacture. Larger companies will operate in all sectors. Their strategy is to diversify activities from relatively specialised bases, for example, a base centred on a product or service offered, or, more frequently, defined by the geographical area in which the company operates. As firms seek to grow the most common diversification has followed the trend towards backward integration into the production of construction materials and forward integration into property development. This trend is coupled with diversification outside the UK by construction companies setting up overseas operations in former Empire territories and in moves to areas such as the Middle East, Far East and North America. Over the period

1986–1998 the value of contracts obtained by UK-owned construction firms operating in the Middle East increased by 47.5%. The Middle East has traditionally been seen as a major location for UK construction firms to move into and this increase in value of contracts obtained is healthy. However other areas of the world are also receiving attention. The value of contracts obtained by UK-owned construction firms in the Far East over the same period increased by an incredible 498% while the value of contracts obtained by UK-owned construction firms operating in North America increased by 152.6%. The role of international markets for construction is discussed in Chapter 6.

The professions

In common with contractors the professions associated with the construction industry may be seen as 'generalists', in that in order to develop and retain flexibility in the face of changing market and economic hazards, they tend to adopt a generalist attitude to their work. Thus architects tend not to specialise in one particular building type or method of construction. This means that most firms are prepared to tackle most, though not necessarily all, building problems within their particular resources. The strategic choices which the professions need to make are therefore related to their markets and most, like contractors, have chosen flexibility as a mechanism for survival and growth.

This concept of the professions as generalists relates to the product – the building being built – rather than to the service being provided to clients. The professions rely on gradations of specialisms with structural or services engineers providing specialist design facilities and quantity surveyors providing expertise in financial management of projects. These are the traditional professions but professions related to the control of particular aspects of the construction process are also involved. For example, professional project management firms who are independent of the contractors and sub-contractors actually executing the work provide specialist management for clients. More recently firms involved in programming the whole of the construction process, from inception to completion, have come into being. This development resulted from the identification of a gap in the market: architects (or other designers) control quality and quantity surveyors control costs but there was no specialist to control 'time'. As professional boundaries change new opportunities emerge and contract claims specialists, quality control experts, building envelope engineers (engineers who are commissioned to design cladding systems for the outsides of buildings) and other professionals currently on the margins of the construction process are redrawing the contours of the responsibilities of the professionals. A strategic consequence of this is the potential for discovering and exploiting new areas in the market.

The consultants

It is not appropriate here to describe the traditional role of designers; this section merely seeks to identify the structure of the professions. The organisation of a design practice will have implications for the strategy of the organisation for a large proportion of designers in private architectural practice are based in one-person or small practices. Whilst the work undertaken by private practices varies with the size and turnover of the practice certain generalisations may be made:

(1) The smaller practices deal mainly with private individual clients or with larger clients requiring small-scale works;
(2) The larger practices are able to deal with corporate clients whether the work is public or private in nature. Commercial and industrial projects feature heavily in the portfolio of buildings designed and major feasibility studies frequently appear in the portfolios of many larger practices.

Important changes have taken place to the strategic role of the architect in the construction process during the 1990s. Perceptions of the architect's role may have resulted in changes which have emphasised the design aspect and diminished the project management role. Certainly new procurement methods such as management contracting and construction management, project management, prime contracting and even design and build methods have imposed changes on the way architects organise their practices.

Also, within the last two decades there have been a number of reports and research studies that have addressed the roles, responsibilities and management issues which face architects and surveyors (e.g. Avis & Gibson [11], Hillier [12], Kelly & Male [13], MAC [14], Male [15], Male & Kelly [16], Male [17], RICS [18], PRS [19] and RIBA [20]).

At a societal and institutional level, a series of important events occurred that shaped the future of the architectural and surveying professions. The investigations of the Monopolies and Mergers Commission into the provision of services by architects and quantity surveyors instigated the procurement of professional services through competition on price. Also, the client-focused and contracting-focused wings of the surveying profession were brought together through the merger between the Institute of Quantity Surveyors and Royal Institution of Chartered Surveyors. There was also a redefinition of roles within professional groupings brought about by new methods of procurement and forms of contract and also the option existed for architects or quantity surveyors to be able to hold directorships in contracting companies.

Architectural practices during the 1970s shifted in size either towards the small or large end of the spectrum. This is also a characteristic of the surveying profession and the smaller architectural practices were found to have up to three times as high a proportion of private sector clients as larger practices, although the latter were found to have more stable clientele. Many architectural practices specialised involuntarily in particular building types as a result of relationships with particular clients. The tendency to concentrate on one particular building type was found to be more pronounced in large and small practices than in those of medium size. Specialisation can limit the potential of a practice to adapt to changing patterns of demand. Also, when dividing a project into strategic (dealing with the client end) and tactical (the construction end) areas, as firm size increases there is a tendency for architects to be more involved in the strategic end of projects. The role of the architectural technician had become well established, unlike the technician's role in quantity surveying which is less well defined by career routes and qualification. The architectural technician can belong to the British Institute for Architectural Technicians whilst no comparable body is available for quantity surveyors.

During the 1980s clients believed that architects should accept responsibility for the full range of normal and supplementary services and direct and co-ordinate all consultants. Clients also believed that architects had an inability to respond to changes or recognise their needs. The primary threat to the architect in all areas, except for design, was seen as coming from the chartered surveyor.

For the surveying profession as a whole during the 1980s, the markets served became more diverse and clients changed their buying behaviour with respect to surveyors' services. Additionally, there was increased competition, not only between surveying firms but also with other professions. Surveyors' markets were characterised by a transition from growth to maturity and from regulation to deregulation. Large quantity surveying practices offered a relatively narrow range of services. It was suggested that such practices should broaden their scope of service provision or be managed as decentralised entrepreneurial units. Additionally, medium-sized general practice firms had been advised to opt for growth, either through internal expansion or acquisition. The quantity surveying profession offered many more services than just preparation of tender documentation and post contract services, the traditional core business of quantity surveying. However, the core business activity still accounted for approximately 50–55% of workload. In general, it was the leading edge of the quantity surveying profession, some 20% of practices, that were involved in the provision of services outside the core business. The addition of new services, such as project management and value management, were taken on board by these practices. The use of sub-contracting within quantity surveying became more prevalent and it was contended that this could force changes in the 'form' that the profession takes and determine future developments in the control of the profession.

Within the majority of general practice firms, research findings indicated that they were not concerned with a future orientation. All types and sizes of firms had difficulty in implementing strategic planning partly because of an inability to take a broad view of the firm and its markets. However, operational management appeared well developed, through the use of regular reporting procedures and meetings. It was also found that in general practice surveying, 65% of practices' income came from existing clients but they were increasingly exercising choice and widening their service procurement options. It was predicted that competition among architects, surveyors and engineers would occur increasingly at the boundaries of a profession's skills and knowledge base, where services are prone to develop and hence, overlap.

In combining the empirical evidence from architecture and surveying the following picture emerges. A restructuring, similar to that in contracting occurred in the size of consultancy firms, namely, a polarisation towards small or large organisations. Firms were having to respond to a diversity of market and client types. However, due to the restructuring processes within professional markets, larger practices had a more stable clientele with the greatest variability being experienced in the small-firm sector. Firm size also had an impact on the form of pressure exerted by clients. As firm size increased professionals became more involved in the strategic end of projects and had more interaction with clients. Empirical evidence from the late 1970s indicated that due to client demands specialisation by project type probably occurred in both large and small practices but not in medium-sized firms.

Historically, client pressures appeared to be forcing consultancies towards niche strategies although with the introduction of fee bidding this process may well have been reversed. Fee bidding may force some consultancies towards a generalist orientation. Evidence from both the surveying and architectural professions has indicated that technicians were widely used in the production of design information and that these staff grew in numbers as firm size increased. For surveying, interdisciplinary competition among firms was likely, leading to a redefinition or virtual disappearance of boundaries between divisions of surveyors. Innovation occurred in the leading edge of the professions, some 20% of practices. This indicated an increased need for flexibility. Surveyors were turning to sub-contracting work to other practices to even out workload problems. Additionally, surveying firms appeared to be adept at operational management but not at strategic management and based on this empirical evidence, the same could be postulated for architectural firms. In architecture, competition in the main was seen to come from the surveying profession, except in the area of design, and as such skill substitution between the professions became apparent in many areas. Thus two forces, client procurement practices and competitive pressures, were acting in combination to force architects towards the design end of the building process. Much the same argument could be put forward for consultant engineers, from whom quantity surveyors were increasingly taking over the financial control aspects of engineering projects.

The picture that has emerged from this analysis is that the strategic decision-makers of firms in the construction industry were facing different types of pressure but moving towards adopting similar strategies to deal with those pressures. Those in contracting, traditionally acting in a competitive environment, were operating different types of diversification strategy in response to historical changes in their business environments. The professions were operating within business environments characterised by higher levels of competition, not only between professions but also within professions. Therefore, strategists within professional consultancy firms were being forced to make strategic decisions not only about how to cope with competition but also about how to respond to a business environment requiring diversification strategies. They were considering boundary spanning activities, acquisitions and mergers, and sub-contracting. The business environments of these firms required the development of strategies and management skills previously characteristic of only a contracting firm. Underpinning this analysis is the requirement of strategists in all types of business to handle change effectively.

Clients of the surveying profession blurred the distinction between professional services, based on formal learning, and commercial services based on the use of experience and therefore not requiring expertise in order to sell them. Architects, traditionally operating within a lead consultant role, were increasingly seen by clients solely as designers rather than as designers and managers of the building process. Weaknesses in the surveyors' position stemmed from an over concern with technical skills at the expense of business and commercial skills. This latter charge has also been levelled at architects. Clients are more often seeking 'one stop shopping' for the purchase of their construction services and so architects have combined, formally or informally, to provide multi-discipline practices which encompass the full range of expertise necessary for the erection of a building.

The changes in the way buildings are procured have taken place against a backcloth of greater commercial awareness amongst architectural practices. As architects move away from their position as 'an occupation possessing a skilled intellectual technique' (Kaye [21]) set within the framework of a voluntary association (RIBA) with its code of conduct, they are offered a new range of strategic choices. Modern architectural practices tend to be corporate identities and require strategic planning to ensure their survival and growth.

Civil, structural and services engineers are in a similar position to architects. However, the professional associations for engineers are not as involved with professional firms as are RIBA or RICS. The principal vehicle for policing consulting engineers' conduct is the Association of Consulting Engineers (ACE) which lays down broad guidelines for conduct and fees, etc. Whilst the engineering professions have not experienced the plethora of changes which have shaped the architects' role they have to operate in the same construction environment. Therefore engineering consultancies have sought to combine with others to enable them to survive and grow.

This redirection of the professions into business entities has meant that several partnerships have become public limited companies and some have sought listings on the Alternative Investment Market (a second division stock exchange listing) and the largest firms have achieved a full stock exchange listing. These events presage a new interest in corporate planning by the construction consultant who will look to take stock of business opportunities within the construction environment.

References

[1] Robbins, S. (1983) *Organisation Theory: The Structure and Design of Organisations.* Prentice Hall, Englewood Cliffs, NJ.

[2] Jauch, L.R. & Glauck, W.F. (1988) *Business Policy and Strategic Management, 4th edition.* McGraw-Hill, Singapore.

[3] Johnson, G. & Scholes, L. (1988) *Exploring Corporate Strategy, 2nd Edition.* Prentice Hall, Hemel Hempstead.

[4] Lansley, P. (1987) Corporate strategy and survival in the UK construction industry. *Journal of Construction Management and Economics.* **5**, No. 2: 141–155.

[5] Ball, M. (1988) *Rebuilding Construction: Economic Change in the British Construction Industry.* Routledge, London.

[6] Lansley, P., Quince, T., Lea, E. (1979) *Flexibility and Efficiency in Construction Management, Final Report.* Building Industry Group, Ashridge Management College, Amersham, Bucks.

[7] Newcombe, R. (1976) The evolution and structure of the construction firm. Unpublished MSc thesis, University College London.

[8] Fleming, M. (1988) 'Construction'. In *The Structure of British Industry,* P. Johnson (ed.). Unwin Hyman, London.

[9] Hillebrandt, P.J. (1984) *Analysis of the British Construction Industry.* Macmillan, London.

[10] Catherwood, F. (1966) Development and organisation of Richard Costain. In: *Business Growth,* R. Edwards & H. Townsend (eds). pp. 271–286. Macmillan, London.

[11] Avis, M. & Gibson, V. (1987) *The Management of General Practice Surveying Firms.* Research papers in Land Management and Development - Management, No. 1, University of Reading.

[12] Hillier, W. (1979) *The Structure of the Profession.* Unpublished report for the Royal Institute of British Architects. Bartlett School of Architecture and Planning, University College, London.

[13] Kelly, J. & Male, S. (1987) *A Study of Value Engineering and Quantity Surveying Practice, Final Report.* Quantity Surveying Division, Royal Institute of Chartered Surveyors, London.

[14] MAC (1985) *Competition and the Chartered Surveyor. Changing Client Demand for the Services of the Chartered Surveyor.* Report by Management Analysis Centre for the Royal Institution of Chartered Surveyors, London.

[15] Male, S. (1984) A critical investigation of professionalism in quantity surveying. Unpublished PhD thesis, Heriot-Watt University, Edinburgh.

[16] Male, S. & Kelly, J. (1989) The organisational respondes of two public sector client bodies in Canada and the implementation process of value management: lessons for the UK construction industry. *Construction Management and Economics,* **7**, No. 3: 203–216.

[17] Male, S. (1990) Professional authority, power and emerging forms of profession in quantity surveying. *Construction Management and Economics,* **8**, No. 2: 191–204.

[18] RICS (1984) *A Study of Quantity Surveying Practice and Client Demand. Report for the Quantity Surveyors Division.* Royal Institution of Chartered Surveyors, Surveyor's Publications, London.

[19] PRS (1987) *The Architect in a Competitive Market. Report by Property Research Services*

for the Cities of London and Westminster Society of Architects and the London Region of the Royal Institute of British Architects. RIBA, London.

[20] RIBA (1992) *A Study of the Profession.* RIBA, London.

[21] Kaye, B. (1960) *The Development of the Architectural Professions in Britain.* Allen & Unwin, London.

4 Clients, constructors and competencies

Introduction

Having documented the changes taking place in the construction industry markets, this chapter sets these markets in the context of the wider business environment for construction. Construction organisations operating within the industry meet the needs of a diverse range of clients, ranging from individuals through to large multinational companies. Construction clients can also operate within either the private or public sectors. The distinction is becoming blurred with the privatisation of utilities often seen historically as providing a public 'good' or through the requirements of new procurement routes such as the Private Finance Initiative, (PFI). Under PFI private sector consortia will finance, design, build and operate public sector facilities for periods of between twenty and thirty years. Within these privately funded and operated facilities, public sector services are now being delivered; for example, hospitals, schools, the courts, police establishments, etc. Clients also vary in their degree of knowledge about how the industry functions, and they may be regular or irregular procurers of facilities, approaching the industry frequently, once only or on an intermittent basis. Finally, UK construction companies and consultancy firms can operate in local, regional, national or global markets.

This chapter sets out the context within which strategic management decisions are to be made. It introduces the concepts of the product and project life cycles and how the industry has evolved. It deals with the business environment in construction, introducing the ideas of the general and task environments, then develops this further by exploring the nature of the industry and market in construction. Finally, the chapter explores construction as a fragmented and hierarchically structured industry and draws the themes of the chapter together into a concluding section.

The concepts of product and project life cycles in construction

The concept of the product life cycle plays a fundamental role in strategic management and this will be explored in the context of construction. The product life cycle is, however, not without its critics when applied to all products.

In simple terms the idea behind the concept is that industries and products go through stages, in a life cycle. The stage of this cycle at which an industry or product is at a given time has important implications for decision-makers. Four life cycle stages have been identified. The development stage is where one or a few firms introduce a new product, risks will be high and there will be high start-up costs in relation to sales. The second stage is the growth stage, where sales and profits increase and prices reduce. Price reduction is, in part, due to the combined effect of competition among firms and cost reductions. The third stage is maturity where sales growth is at a slower pace than previously and profits and prices reduce due to increased capacity. The final stage is decline. Market demand has to a large extent been satisfied and there is over capacity in the industry. Prices will remain stable or begin to fall along with profits and the particular product will probably change into a loss maker.

In construction there is a service product, for example, design and build or management contracting and an end-product, the completed structure or facility delivered to the client. The service-product will have a life cycle similar to that outlined above. There are a variety of possible service-product inputs, related to organising the project delivery process, that can achieve the delivery of the end-product. These would include, for example, traditional contracting, design and build, management contracting, etc.

Built or modified structures, realised through a project-based manufacturing process, are generally one-off and have a tendency towards uniqueness. The term normally associated with the end-product in construction, due to its project-based characteristics, is that of the project life cycle. The typical project life cycle is concept, design, construction, handover and commissioning, operation and disposal. However, as the eLSEwise project [1] on large scale engineering has established, this confuses project process life cycles and end-product life cycles from the client's perspective. In construction, the client is interested in a completed facility of some kind, a physical asset, that permits ongoing business functions to take place. Operation and Disposal are concerned with the end-product of the project delivery process, the built structure. The project process only comprises a life cycle of stages from concept to handover and commissioning. Procurement strategies provide different mechanisms to organise managerially and contractually the project process as a service-product in construction. Therefore, design and build, management contracting and other forms of procurement service-products will go through the different product life cycles of their own, as identified above. The end-product in construction will go through the project and end-product life cycles. Some of these end products will be closer to a manufacturer's production model, others will be akin to bespoke craft production.

The strategic concepts of industry and market in construction

The structure of an industry directly impacts the nature of competition between firms in that industry and the competitive strategies available to them [2]. The construction industry in the United Kingdom has traditionally been viewed as being sub-divided into the civil engineering and building industries. However, by their nature, many construction contracts will involve both aspects of building and civils work. Developing a competitive strategy in construction requires the arena within which competition takes place to be defined and normally encompasses the concepts of an industry and a market.

Analytically, an *industry* is a supply side concept. It is an arbitrary boundary within which firms compete with each other to produce related or similar products. There are five major forces determining industry structure and these jointly establish the profit potential in an industry. An awareness of their joint influences is important for understanding the competitive environment facing the individual firm. Figure 4.1 sets out these forces diagrammatically. Described briefly, *buyers* and *suppliers* have similar effects in that if they are powerful, profit margins can be pushed down. Buyers (or clients) as a group are particularly important in construction since their advisers, in the form of the construction professions, can dictate the rules of competition for contractors, especially through the choice of procurement path. Procurement strategies are expanded on further below. In the case of suppliers, construction is a highly interconnected industry through materials inputs from many other industries. Therefore, where industrial concentration in other industries may be high the opportunity for suppliers to influence input costs could be considerable. *Threat of entry* is concerned with the likelihood of new competitors entering the industry. This is dependent on the presence or absence of entry barriers. It is

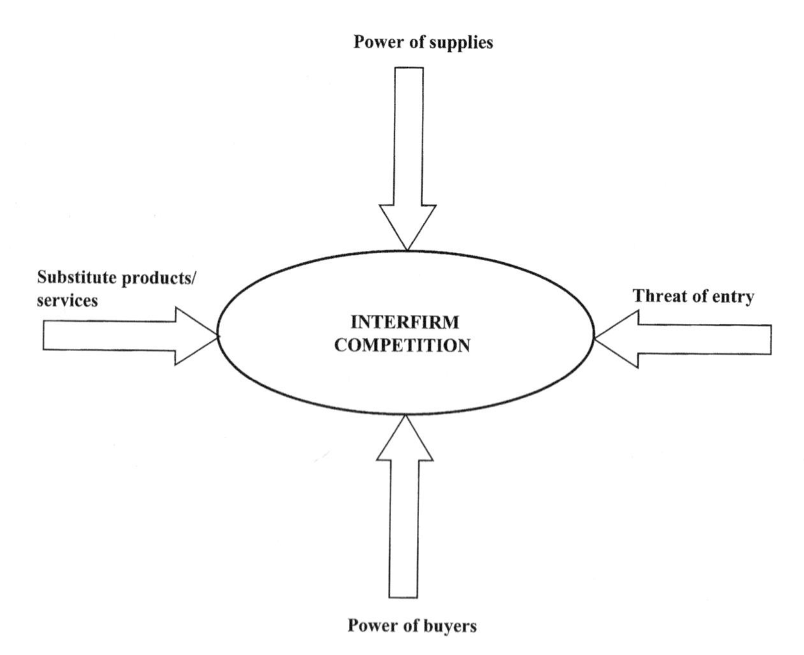

Fig. 4.1 The forces shaping industry structure. Source: Adapted from Porter [2].

often claimed that entry and exit barriers to the construction industry are low. However, others have suggested that there are subtle forms of entry barriers present in construction that have no direct counterpart in manufacturing. This stems directly from the heterogeneity and characteristics of the industry and the nature of the markets within it. Entry and exit barriers are discussed further below. *Substitute products or services* are not an easy concept to apply in the construction industry. The critical issue in deciding what is or is not a substitute, according to Porter [2], is that the substitute must undertake the same function. Substitute products and services in construction are elaborated below. Finally, the *extent of competitive rivalry* is determined by the degree of mutual dependency or interaction between competitors and the likelihood of this setting off retaliatory strategic moves between them.

A *market* is a demand side concept, having both an economic and social dimension. It organises an exchange relationship between buyers and sellers of a commodity to determine price [3].

Kay [4] provides a useful distinction between markets and industries and also introduces the term 'strategic groups' into the equation:

- A market is defined by demand conditions, is based on consumer needs and is characterised by the 'law of one price', and is bounded by the ability of the consumer to substitute one product for another.
- An industry is determined by supply conditions, is based on production technology and is defined by the markets chosen by firms. Industries are determined by the manner in which production is organised.

A strategic group is defined by the strategic choices made by firms in terms of who they decide to compete against, it is based on a firm's distinctive capabilities and market positioning, with group membership determined subjectively by management within different competing firms. Strategic groups in the construction industry are discussed further in Chapter 5.

To Kay, the important issue for the firm is its choice of markets in terms of product and geography. Membership of a strategic group and industry follows from that choice. Hence, industries are relatively stable and are based around outputs from production capabilities whilst markets are transient, are determined by consumer needs and are bounded by substitute products and the price consumers are prepared to pay for the outputs.

Market structure, price determination and competition in construction

There are two distinct types of market structure in construction with different economic forces operating in each type [5]:

(1) Contracting, which involves a company in constructing a facility to a customised design where the roles and responsibilities of the constituent

parties are contractually defined. The method of price determination is the reverse of manufacturing in that the contractor determines price prior to production. Under this form of market structure built structures, as an end-product, are pre-demanded by the client.

(2) Speculative construction, which involves anticipating, responding to or creating demand, and more typical of a traditional manufacturing approach. A typical example in construction is speculative house building.

Different skills are required for each type of market structure. In the first instance the emphasis is more on managerial and technical skills. In the second, entrepreneurial activity involves market forecasting of a different type, with market research, the assembly of financial packages and land banks.

Procurement and tendering strategies in construction to contract

Procurement strategies are the options available to the client in the market place for obtaining a facility through a managerial and administrative framework that is set up to handle the project process on behalf of the client. There are six generally recognised procurement options available to the client that integrate design and on-site production more or less effectively in different ways:

- Traditional, separating design from construction;
- Management contracting;
- Construction management
- Project management;
- Design and construct;
- Develop and construct;
- Separate contracts;
- British Property Federation.

A further four can also now be added to the list, namely,

- Turnkey, which is similar to design and build but will include full fit-out and may include financing;
- Build, Operate and Transfer (BOT);
- Private Finance Initiative, essentially finance, design, build, operate and transfer (FDBOT);
- Prime Contracting, where a contractor leads supply chain management for the client.

These twelve procurement options set up a range of different organisational, managerial, and administrative and risk relationships between the client, the client's advisers and the contractor. Additionally, in contracting each procurement strategy also modifies the market structure faced by the contractor.

Tendering strategy is the market mechanism for selecting, choosing and appointing a contractor. There are two main approaches to contractor selection, with a series of sub-options:

(1) By negotiation: where only one contractor is involved.
(2) By competition, with sub-sets as follows:
  - Open competition, where any number of contractors can compete;
  - Selective competition (single stage), with normally between four and six contractors in competition, based on a pre-qualification process to ensure they have the competence to undertake the work;
  - Two-stage tendering: combining selective competition in the first stage and then negotiation in the second stage;
  - Serial/continuity contracts: combining competition initially and then negotiation for a series of similar projects. This type of contract facilitates project learning but has considerable risks attached during periods of high or rampant inflation.

Gasttorna & Walters [6] introduce a useful analytical concept for service provision, namely, a *qualifying level of service* which is available from all serious competitors and represents the basic need to survive and remain in competition. The *determining level of service* is the package that is unique to the seller, adds value to the customer and provides competitive advantage. In applying these concepts to construction, any form of pre-qualification involves assessment of a contractor's ability and competence to undertake the work. Serious competitors are those that are considered to have the competency to undertake the work in hand based on expertise combined with company reputation. Once pre-qualification has taken place the assumption is that all those selected are serious competitors and hence competent to undertake the work. Pre-qualification determines the qualifying level and consequently expertise and company reputation cease to be forms of competitive advantage. The lowest price culture that dominates the construction industry currently means essentially that price is the determining level of service. This axiom is being challenged by the best value model of awarding public sector contracts. In such arrangements soft issues such as relationships with the parties in the content and attitudes to, say, partnering are likely to be as important, if not more so, as price.

Contractual arrangements

Contractual arrangements make up the final component locking the market mechanisms procurement and tendering strategies into place. Contractual arrangements provide the legal framework set up to formalise the relationship between the client, the client's advisers and the contractor. Contractual arrangements are intertwined with tendering strategy in determining price and with the method of payment to the contractor. Contractual arrangements can be broken down into the following six categories:

(1) Drawings and specification;
(2) Firm bill of quantities;
(3) Approximate bill of quantities;
(4) Schedule of rates:
  - Standard pre-priced schedules;
  - Specially priced schedules;
(5) Prime cost and cost reimbursable:
  - Cost plus percentage fee;
  - Cost plus fixed fee;
  - Target cost;
(6) The form of contract - standard or purpose written.

The contractual arrangements are essentially the principal mechanism for communicating the client's value requirements to the contractor in a formal manner. They are drawn up normally by the client's consultant advisers.

This sub-section has reviewed the different influences on the operations of markets in construction. The following sub-section discusses the ease or difficulty with which contractors are able to enter or leave a market.

Entry and exit barriers defining construction as a project-based, vertically structured market place

Entry and exit barriers are industry structural concepts related to the ease with which a firm is able to enter or leave an industry. Industries with low entry or exit barriers are easy to enter and leave. The presence of entry and exit barriers can also relate to the levels of profit enjoyed by firms in an industry, with high entry and exit barriers providing a considerable degree of protection to existing firms in an industry.

The construction industry has been quoted as having low entry and exit barriers [7]. This revolves predominantly around its low capital requirements and that 'know-how', the stock-in trade of the contractor, is easily 'poachable' through hiring [8]. However, Male [8] concluded the empirical evidence on the presence of entry and exit barriers in construction is inconclusive. In an attempt to explain the inconclusive results and determine the nature of entry and exit barriers in construction he utilised the structural sources identified by Porter [2] and these are discussed below.

Product differentiation

The product in construction is either a service-product or an end-product comprising a new-build or modified structure through refurbishment or renovation. The product, as a service, embodies reputation and product differentiation in construction occurs through any pre-qualification mechanisms offered in the procurement and tendering strategies adopted by the client. Pre-qualification therefore differentiates one group of contracting companies from another since the pre-qualifiers are seen to have the expertise to carry out a project and, by implication, the non pre-qualifiers do not. Stokes [9] contends, however, that reputation and quality have little impact on the contracting industry for obtaining work. He argues that the final arbiter is price. Sometimes this is not always the case, for example, where the client's consultant advisers may be concerned about the price submitted, if it appears too low. In this instance, reputation is back on the agenda as a differentiator. Reputation, and not lowest prices, can still be a further differentiator through the clear articulation and use of *qualifying* and *determining* levels of service mentioned earlier. It requires the client/consultant advisers to look beyond just the qualifying level of reputation and expertise to get firms onto a pre-qualifying list identifying value-adding activities, normally rooted in competitive advantages.

Male [10] has argued the price submitted differentiates competency in the bidding process. All things being equal, the price submitted is a quantifiable measure, in monetary terms, of the expertise and competency of those involved in the tendering process within a contracting company. This competency is reflected through the adequacy of a firm's market intelligence network; the ability to assess the state of the market, the competition and profit level required, and the tender adjudication process itself prior to submitting a bid. The tender price differentiates in-house expertise (managerial, technical and commercial) between companies in the bidding process.

Differentiation also occurs through the pressure placed on the industry to respond to clients' diverse requirements through new forms of service, such as construction management, design and build and more recently Prime Contracting and PFI. Product differentiation is an important factor within international construction [7] and will be addressed further in Chapter 6.

Capital requirements

Capital requirements can be viewed from a number of perspectives. First, the need to own, invest in and maintain fixed assets. The requirement to own fixed assets in contracting is lower than in other sectors of industry, especially at the lower end of company size. Second, the requirement in contracting to invest in human capital at all levels is high, especially on the more demanding projects, and there is a need for constant attention to training and management development. Historically, the construction industry has been castigated for a poor training and management development record and there is a limit to the extent to which contractors are able to continually 'poach' human resources from other companies in order to circumnavigate the need to continually invest in human resources. Third, as firms in the industry diversify from contracting into, for example, property development, building materials manufacture or mining operations, as many UK contractors have, the extent of physical capital requirements increases due to the changing nature of investment in physical assets and technological requirements. Diversification strategies, to be discussed in more detail later, are now setting up different forms of entry and exit barriers to the industry for construction companies. Fourth, capital requirement (or access to finance) is becoming important for contractor-led PFI consortia and the ability to fund the substantial tendering costs involved as well as the need to convince clients of financial stability. Fifth, diversification into overseas markets also requires considerable capital investment and access to capital to provide financial packages through globalisation/internationalisation.

Switching costs of clients

Clients of construction, as buyers of the industry's services, have a number of options for procuring contractors' services as indicated earlier. There are many different types of clients to construction, from the individual domestic client through to large multinational companies or foreign governments. Clients are also regular and irregular procurers of construction services. Switching costs refer to the cost of moving from one supplier to another. *Switching costs* in construction depend on client type, knowledge of the industry and type of project. In the case of a large complex project, where only a few contractors, and indeed consultants, are capable of undertaking the work, switching costs can be high, especially if it is a negotiated contract [12]. Where there are many firms able to offer the same service, for example, on more straightforward projects, switching costs may be lower. Finally, the requirement to place contracts above a certain size in the European OJEC journal adds a certain rigidity to procurement of services and hence switching costs since the process may take up to three months to complete.

Access to distribution channels

Distribution channels in construction can be of two types, namely, tangible physical or intangible channels. Tangible physical channels are those related to the distribution of manufactured goods where construction companies with, for example, building materials manufacturing subsidiaries, may channel products through the large merchanting retail chains or builders' merchants. More importantly for the focus of this book, intangible distribution channels are those stemming from clients' consultant advisers, who act as institutionalised distribution channels for contractors' services under the Contracting System. Access to these institutionalised distribution channels will require investment in marketing and promotional activities to ensure that a contracting firm's capabilities remain at the forefront of client advisors' minds when it comes to staying on or entering pre-qualification lists. Mergers and acquisitions also allow contractors access to other client and consultant contacts and select tender lists, and hence extend access to consultant advisor distribution channels.

Scale Economies and the Experience Curve

The concept of the experience curve stems from the work of the Boston Consulting Group, which has its critics. The underlying assumption behind the curve is that price levels will be similar for similar products in the same market segment. Products are therefore close substitutes. The difference in the level of profits between companies is therefore dependent on the costs of an individual firm. Thus over time as the total number of units produced increases, the unit cost will decrease. The reasoning behind this is related to the fact that:

- People will learn to do a job more effectively over time;
- Through increasing specialisation and economies of scale, capital costs decrease relative to increasing capacity; and
- Gaining and holding market share is important because of the postulated cost/experience relationship.

The experience curve is not a natural law but requires a concerted effort on the part of managers to reduce costs. It has been argued that economies of scale are not a major issue in construction [7]. However, where they do exist they involve the following [12]:

- Expertise in integrating components of the production process and moving sections of the workforce between projects;
- marketing (acting through the merchant-producer role of the contractor to increase bargaining power with clients and suppliers) and also access to finance;
- functional specialisation by trade, a sub-contracting issue for the main contractor;
- the degree of repetitive work that is allowed through the ability to obtain projects of a similar type, house building being a case in point or projects obtained through serial tendering. This provides economies from learning and is dependent on continuity of work and retaining knowledge in-house and over time. Hiring-in expertise disrupts economies of learning since new management, while having experience of similar project types, will take time to learn and adjust to the systems and procedures in the new firm;
- the diversified construction company is likely to have economies of scale stemming from a manufacturing capability where this type of business operates within its portfolio of subsidiaries.

Construction benefits from economies of scale and learning.

Government policy

This is an important barrier for market entry in international construction. For example, the legal requirement to use local companies or obtain local operating licences are forms of entry barrier [13] [14].

Conclusions on entry and exit barriers in construction

In reviewing and consolidating the above, entry and exit barriers are present in the industry and they can be different from or similar to those in manufacturing depending on the diversified structure of the construction company and the type of market or industry being entered. Entry at the lower end of the contracting industry in terms of firm size is relatively easy. As project size, complexity and technological requirements increase, coupled with any international dimension, there are fewer firms able to undertake such work. Product differentiation stems from first, competencies in managerial and technological know-how, embodied in company reputation and expertise. Second, through access to finance that create barriers to entry for particular types of project and hence arrange contracting into a geographically dispersed project-based market defined by project size and complexity. Third, through a clearly articulated qualifying service and then determining service that provides competitive advantages through adding value to the client. Furthermore, contractors, through diversification, are now involved in different types of production activity requiring different types of capital investment, both human and physical. The diversified construction firm has different types of entry and exit barriers and the empirical evidence appears inconclusive in terms of their effects on profitability at the corporate level.

Conclusion

This chapter has reviewed the evolution of the industry in terms of the contractors and consultant advisors to the client. The business environment in construction has also been reviewed. Within this the importance of demand has been highlighted. Demand plays an important part in determining the nature of the construction industry with its effects being felt mainly in the long term as shifts between market sectors.

A market brings together a buyer with a need for an end-product and a seller who can meet that need with similar or closely related service-products. The latter can satisfy the former's need at a price that is mutually agreed through an economic and social exchange relationship. In construction, the market is client-generated and clients will have different degrees of knowledge of how the industry operates, they may be regular, volume, procuring clients or ad-hoc, generally one-off or intermittent procurers. A market involves an economic exchange relationship to deliver a facility through a project process that requires firms to come together for the benefit of the client, normally through a procurement process that may or may not involve competition of some kind. An industry, on the other hand, provides a longer-term orientation where a company has to work out its corporate/business strategy in order to allow it to compete in particular markets. The five competitive forces determining industry structure that have been identified by Porter set the framework and context within which markets for construction operate. The industry provides the longer-term structural context of competition whereas the market translates this longer-term perspective into a short-term exchange relationship for setting up a geographically located production process. The buyer is prepared to pay a price to the seller in order to have a need satisfied, the buyer and seller are also involved in a value-based relationship focused through the point of economic exchange in a project-based market place.

The concept of the industry and the forces shaping it, as identified by Porter, are not without problems in construction [15]. Substitute products, important for defining market and industry boundaries, are predominantly a buyer issue, focusing on meeting a given need, provided that a seller can offer a similar product able to meet the functional requirements sought by the customer at the right price. As argued in this chapter, it is possible to define the construction industry by the end-product or by different service-products that are concerned with organising the project delivery process as a life cycle. The service-product, from the client's viewpoint can also be characterised by:

(1) The need for single point responsibility. There are a number of options available to the client through choice of procurement route. For example, design and build, prime contracting, PFI, turnkey and consultant-led executive project management are all forms of procurement offering the client single point responsibility for the delivery of a facility. Each could define an industry boundary or segment for competitors in their own right. Functionally they provide the client with single point service responsibility but allocate and manage time, cost quality, risk and the supply chain in different ways.
(2) Bringing on construction expertise earlier in the project delivery process. Management contracting and construction management are distinct service-product types and yet the underlying function is that the contractor is providing management expertise to the client and design team but allocate and manage risk in a different manner.

From the end-product perspective, other forms of substitutes in construction are new build versus refurbishment; repair and maintenance; and renovation. A functional analysis of the end-product substitution process suggests that the underlying purpose is to provide the client with a facility that allows ongoing organisational and business processes to proceed optimally. The solution for the client, therefore, may not be in constructing a new facility but renovating, refurbishing or maintaining an existing facility.

The key issue in distinguishing what is a substitute product, in Porter's [2] or Kay's [4] terms, is to identify the functional usage(s) for the buyer (client) of the end-product, once a decision to build has been taken. The next stage is to determine the level of integration within the project delivery process sought by the client, and subsequently determine the appropriate alternative means of satisfying that function whilst taking account of the management of time, cost, quality, risk and safety through the correct strategic choice of procurement route. This will then determine the competitive arena or market for construction firms. The industry will comprise different strategic groups of firms that have developed organisational structures to meet different client, sectoral workload demands and investment opportunities first, in similar ways within strategic groups and second, differently between strategic groups.

Procurement, tendering and contractual strategies are demand side choices that bind the constituent parties together into the project delivery process and its life cycle. Furthermore, depending on the level of client knowledge of the industry, procurement and tendering strategies are invariably impacted or influenced by design team consultant advisers acting in conjunction with the client – architects, quantity surveyors, project managers and/or engineers. Contractor pre-selection, with final selection undertaken either through negotiation or competition, is generally undertaken by design team consultant advisors who act as a form of distribution channel for contractors' services to the client. Through their advice to clients on procurement, tendering and contractual strategies, construction consultants act as intermediaries in the market exchange relationship between the client and contractor. This 'contracting system' has effectively set up formalised and institutionalised distribution channels for contractors' services. To some extent, procurement options such as 'turnkey projects' modify these relationships and the newer procurement routes such as Prime Contracting and PFI are turning these institutionalised distribution channels on their head and weakening the power of the consultant advisors who are now becoming part of contractors' supply chains.

Male [8] concluded that long-term, strategic thinking is possible in construction and that the indicators are present. Many of the myths surrounding construction in terms of its uniqueness, its long term volatility and highly variable demand, have been challenged empirically although the evidence for short and medium term uncertainty is convincing. The key requirements for strategists in construction are to develop skills that are not internal, production oriented and short-term focused but are strategic, external and long-term focused and involve determining the relationship between the company and its external environment. This is the subject of the next chapter.

References

[1] Zantke, G. & Mangels, B. (1999) Public sector client–private sector project: transferring the state construction administration into private hands. *Engineering, Construction and Architectural Management*, **6**, No. 1, 78–87.
Also, as a wider view of the eLSEwise Project: Male, S.P. & Mitrovic, D. (1999) Trends in world markets and LSE industry, *Engineering Construction and Architectural Management*, **6**, No. 1, pp. 7–20.

[2] Porter, M.F. (1980) *Competitive Strategy: Techniques for Analysing Industries and Competitors*. Free Press, New York.

[3] Hillebrandt, P.M. (1985) *An Analysis of the British Construction Industry*. Macmillan, London.
[4] Kay, J. (1993) *Foundations of Corporate Success*. Oxford University Press, Oxford.
[5] Ball, M. (1988) *Rebuilding Construction. Economic Change and the Construction Industry*. Routledge, London.
[6] Gasttorna, J.L. & Walters, D.W. (1996) *Managing the Supply Chain: A Strategic Perspective*. Macmillan, Basingstoke.
[7] Seymour, H. (1987) *The Multinational Construction Industry*. Croon Helm, London.
[8] Male, S.P. (1991a) Strategic Management in Construction: Conceptual Foundations. In: *Competitive Advantage in Construction*, S.P. Male & R.K. Stocks (eds), pp. 5–44. Butterworth-Heinemann, Oxford.
[9] Stokes, F.H. (1977) *Practical problems in corporate planning: II John Laing*. In: *Corporate Strategy and Planning*, reprinted 1982, B. Taylor & J.R. Sparkes (eds). Heinemann, Oxford.
[10] Male, S.P. (1991) Strategic Management and Competitive Advantage in Construction. In: *Competitive Advantage in Construction*, S.P. Male & K. Stocks (eds), pp. 45–104. Butterworth-Heinemann, Oxford.
[11] Flanagan, R. & Norman, C. (1989) *Pricing Policy*. In: *The Management of Construction Firms. Aspects of Theory*, P.M. Hillebrandt & J. Cannon (eds). Macmillan, Basingstoke.
[12] Ball, M. & Cullen, A. (1980) *Merger and accumulation in the British construction industry 1960–1970*. Birkbeck Discussion Paper No. 73, Birkbeck College, University of London.
[13] Verillo, J., Brazil, S. (1988) In: *The Global Construction Industry: Strategies for Entry, Growth & Survival*, W.P. Strassman & J. Wells (eds). Unwin Hyman, London.
[14] Bennett, J., Flanagan, R., Norman, C. (1987) *Capital and Counties Report: Japanese Construction Industry*. Centre for Strategic Studies, University of Reading, Reading.
[15] Langford, D. & Male, S. (1991) *Strategic Management in Construction*. Gower, Aldershot.

B Concepts of strategic management

5 The firm and the strategic management process

Introduction

This chapter is divided up into a series of sections that commence with a series of models of the firm and then progress to describe some of the core concepts of strategy and strategic management. The chapter synthesises information from the general business literature with that specific to construction. The chapter lays the foundations for subsequent chapters dealing with international business strategy in construction, the strategic management of construction firms and consultancy organisations and also supply chain management in construction.

Diagnostic models of a firm

The term 'firm' is used here to represent a social organisation with one of its main objectives being making a profit. Firms normally have some form of vertical structure, or hierarchy, and operate within an external business environment (see Figure 5.1).

The external business environment of a firm, discussed in the previous chapter, can be viewed as an envelope that surrounds the firm and impacts it through different types of pressure. The firm is delineated from the external environment by a boundary. The firm as an economic entity and social system can be conceived as having a moveable or permeable boundary because people from within it are in constant interaction with others from organisations outside the boundary, either by telephone or through meetings, etc. For an architectural practice, the production process is, in part, concerned with the transformation of ideas and information from the client into outputs in the form of a set of hard-copy of CAD drawings for the contractor.

In considering the hierarchy of a firm, Figure 5.1 identifies different levels of management. Each level within the firm has different properties in terms of the level of uncertainty encountered and also requires different skills of managers. The *institutional* or *strategic level* is concerned with adapting the organisation to the external environment. At the strategic level managers are concerned with those issues that relate the firm to the environment. The question here is one of survival and adaptation to the pressures of the external environment. Uncertainty is at its highest and information is less, is ambiguous and unstructured. Managers are concerned with the longer term and there is a requirement for conceptual and judgemental skills and the ability to handle, analyse and

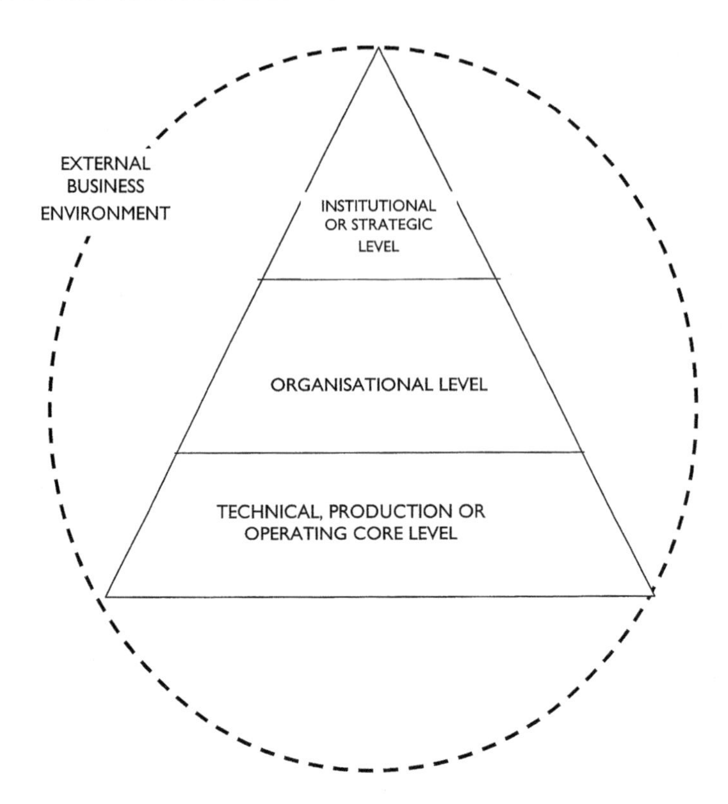

Fig. 5.1 A simple model of a firm. Source: Adapted from Langford & Male [1].

reformulate unstructured information received from the environment and internally from within the firm. Typical examples in construction would be director level in contractors and partner level in consultancies. The development of skills and concern with the firm's requirements at this level have tended to be neglected in the construction industry.

The focus of managers at the *organisational level* is integration of lateral and vertical relationships in the firm. The concern at this level is one of mediation between the strategic level and the technical level as well as maintaining lateral relationships between departments, other teams or other functions. There is less emphasis at this level on technical skills but more on political and organisational skills such as the ability to handle people, organisational systems, procedures and controls. Uncertainty at this level has decreased. Managers need a task focus and capabilities requiring negotiation, compromise and political astuteness. Typical examples of the organisational level in the construction industry would be contracts managers in contracting companies or team leaders in consultancy organisations. Managers at this level need to operate in two different time frames, the short term when dealing with the production level and the longer term when dealing with the strategic level.

Finally, the *technical, production or operating core level* is concerned with transforming inputs from the environment into outputs which can subsequently be sold in the market place. The production level is task oriented, uncertainty is relatively low and routines can be developed for handling repetitive tasks. The production level is concerned very much with the present and getting the job done. Managers require a task orientation, technical skills and a short-term time horizon. This would correspond to the site agent level in a construction company, draughtsman level in an architectural or engineering consultancy or a surveyor in a quantity surveying practice who is involved in producing tender information.

Within the management hierarchy different types of decisions and decision-making strategies will be present. Broadly, there are three approaches to decision making [2]:

(1) The rational-analytic approach, where decisions are viewed as being analytical, conscious and rational and where the decision-maker considers all available alternatives systematically. This is the traditional, economist's view of decision-making.
(2) The intuitive-emotional approach, where a number of alternatives are considered but unconscious processes are used by the decision-maker to save time in the decision-making process.
(3) The political-behavioural approach, where a decision-maker may be faced with competing pressures and decision outcomes are a compromise through negotiation and mutual adjustment. The political ramifications of implementation are also taken into account as part of the decision making process.

During the decision-making process, people will not have complete access to all information to make an informed decision. Simon [3] has demonstrated that people will decide on the first acceptable alternative, namely, the one that is good enough, rather than take the rational-analytic approach. Simon used the term bounded rationality to describe this as satisficing rather than decision optimising behaviour.

In addition to different approaches to decision making within the management hierarchy there will also be a number of different types of decisions, for example Feldman & Arnold [4] have identified four types of decisions within an organisation.

(1) Personal decisions, which are those made by individuals that can impact their personal lives, career moves being a prime example.
(2) Organisational decisions, which are those concerned with on-going processes within the organisation. Examples would be decisions about procedures, rules, policies and budgets.
(3) Programmed decisions, which stem from the routine, repetitive and frequently occurring situations within organisations and assist in developing standard procedures.

(4) Non-programmed decisions, which are decisions occurring during situations that are ambiguous, unstructured, novel and/or complex and where there are no guidelines from policies, procedures or routines to assist the decision-maker. This requires a response from the decision-maker that is problem-solving or problem-seeking and with a high degree of insight. In the simple model identified in Figure 5.1, these types of decisions are common at the institutional or strategic level.

Ansoff [5] articulated three types of decisions in organisations:

- Operating decisions, dealing with transforming inputs into outputs. The bulk of a firm's time will be devoted to such activities and these types of decisions are concerned with the resource conversion process. Examples, in construction companies include [6]:
  - pricing (bidding);
  - establishing a market strategy (which project type/service);
  - production scheduling (site planning);
  - budgetary allocations among functions (departmental budgets, project budgets).
- Administrative decisions, concerned with organisational structuring and resource allocation throughout the structure of the firm.
- Strategic decisions, which relate the firm to its business environment. Their effect is diffused throughout the organisation over time and therefore has an impact on the previous two forms of decisions. Strategic decisions are externally focused rather than inwardly within the firm and are to do with deciding what business the company is in currently and what it should be in.

It is less clear-cut where, within the hierarchy, administrative and operational decisions will be made, much will depend on the size of firm. However, as a guide, once the strategic level has set out the framework for strategy development and implementation within the firm, administrative decisions are more likely to comprise the remit of the organisational level whilst operational decisions are more likely to form the remit of the production level.

In summary, this section has presented a simple model of a firm with different layers of managers requiring different skills and time horizons, and making different types of decisions. The firm operates in an environment to which it has to adapt and decision-making in firms is influenced by many factors. People in organisations use different decision strategies at different times but, due to bounded rationality, satisfice rather than optimise. The section has also presented a range of different approaches and types of decisions that will be taken in a firm.

The following sections deal in more detail with the nature of firms as construction organisations by presenting a more complex model of a firm.

A firm as a more complex organisation

Male [7] defines an organisation as an ongoing, goal-directed undertaking comprised of people whose activities are co-ordinated through some form of organisational structure. The organisational structure co-ordinates the division of labour that has been allotted by management to perform tasks or responsibilities [5] and has two basic functions. First, it is concerned with reducing the variability present in human behaviour within the firm to enable the organisation as a social and economic entity to have a common purpose. Second, organisational structure is the context within which power is exercised, decisions made and information flows take place [8]. Mintzberg proposes that co-ordination within the organisational structure can be achieved in five basic ways which are not necessarily mutually exclusive [8]:

Mutual adjustment, operating through informal communication, and working best in simple and complex organisational situations.
Direct supervision, involving one individual having responsibility for the work of others.
Standardisation of work processes, involving specifying or programming the content of work.
Standardisation of outputs, involving specifying the results of the work to be achieved. Mintzberg quotes the example of standardising the dimensions of a product or the performance required.
Standardisation of skills (and knowledge), involving specifying the training required to perform the work or task.

Robbins [9] has identified three basic components to organisational structure, namely:

- *Complexity,* a structural concept relating the extent to which the organisation differentiates activities horizontally, vertically and spatially. *Horizontal differentiation,* is the extent to which tasks are sub-divided among organisational members such that they can be allocated to specialists or non-specialists. Horizontal differentiation also concerns the degree of specialisation within an organisation. *Vertical differentiation* is the number of levels in the organisational hierarchy. *Spatial dispersion* is related to either vertical or horizontal differentiation, and refers to the extent to which activities or personnel are dispersed spatially by separating power centres or tasks. Using these definitions of structure, contracting companies are organisationally complex. They will have estimating, buying, surveying and contracts management departments, differentiating the structure horizontally. Vertical differentiation occurs through having a managing director, functional directors, regional directors, contracts managers and site agents whilst spatial dispersion occurs through geographically dispersed regional operating units (subsidiaries) and also construction sites that report to a region's operating unit.

- *Formalisation* is concerned with the extent to which codes of conduct or the norms of an organisation are explicitly known amongst its members. Often formalisation is seen as purely the written rules and procedures operating within the organisation. This is only one aspect of formalisation since unwritten norms and standards can be as potent in controlling human behaviour as those in writing. Mintzberg [8] views formalisation and training (in order to standardise skills) as substitutes since they are both methods of co-ordination.
- *Centralisation* refers to the degree to which power is centralised or concentrated within the hands of a few people, units or departments within an organisation. The extent of centralisation is also a measure of trust within a firm and the extent to which individuals are allowed a wide degree of latitude in making decisions. Power exists along a continuum from highly centralised to highly decentralised. Decentralisation has vertical and horizontal components. Organisations are vertically decentralised when power is formally distributed down the management line hierarchy. Organisations are horizontally decentralised when decision-making power rests mainly outside this hierarchy but lies within the staff function, normally comprising specialist experts, such as accountants.

Finally, the firm, as an organisation, can be viewed either from a structural perspective comprising those aspects identified above or from a situational perspective comprising a network or system of organisational roles. A role perspective of an organisation starts from the premise that an individual in a firm has expectations, perceptions or beliefs about how they behave in a social situation. Equally, not only does the individual holds these views about themselves but also how other people behave in social situations. A role analysis of a firm holds that these two perspectives are held by an individual to create a view of a firm which is a social organisation where the individual has a role [10].

Hunt [11] has brought together this situational view of an organisation into an analytical model. He sees the *formal structure* concerned with first, the co-ordination and control of people's activities within the organisation; second, casting individuals into jobs which are separated by status, power and authority for making decisions; third, the rules, procedures, guidelines and records that control an individuals' behaviour. The formal structure represents the codified history and decisions of the organisation and changing it is difficult and takes time. The *informal structure* is based on emotion and the need for social relationships and the informal, interpersonal relationships that develop within any organisation. The informal structure continually changes, is unstable and is ambiguous and has a number of positive and potentially negative aspects associated with it. Relationships operate outside the formal structure and often bypass it but the informal structure provides flexibility and is the 'glue' that binds the organisation together. The *technical and (physical) system* describes all of the hand tools, machinery, computers, buildings and the internal physical environment of the workplace that impact potentially on the social system. Hunt makes a clear distinction between the technical system and technology. He considers technology as embodying aspects of social adaptation, knowledge and skills that relate to the physical instruments used by the organisation. Hunt sees technology referring to a combination of inputs with:

- the machines of the technical system;
- the members controlling them;
- the knowledge of the designers of the process.

Finally, analysis of a firm's technical system can be undertaken in three ways. First, in terms of flexibility of the technical system and the degree of choice allowed by those operating the technical system. Second, technical complexity required by staff coupled with the number of actions involved in a process and third, the complexity of the technology. Technology therefore involves aspect of organisational structure and process and has a total system impact. As a consequence it is important, but is often the slowest thing to change in an organisation.

The last analytical component of the organisational role system is the individual variable, referring to the impact that an individual's personality, ambitions, goals, desires and needs can have on an organisation. The organisation, as an on-going social organism has to adapt to change and has to consider the similarities and differences between individuals and the fact that an individual's personal objectives may not always align with those of the firm.

Finally, the impact of the external environment, described in the previous chapter, has to be considered on the role system. Managers appear to make decisions based on their perceptions of the external environment, often with little relationship empirically between measures of the objective (or actual) environment and the perceived environment used for decision-making [12]. It is managers perceptions that create the environment to which the firm responds and these, in turn, will determine the firm's structure and the assessment of uncertainty it faces.

This section has described two models of a firm, a simple and more complex model. The next section reviews research on the concept of strategy.

The concept of strategy

Management theorists agree that strategy deals with the *means* an organisation uses to meet *ends*. Traditionally, it has been seen as concerning the achievement of long-term objectives [7]. Strategy has also been viewed as a set of rules to guide decision-makers about organisational behaviour. Strategies may be explicit or implicit, may remain within the senior management team or pervade an organisation and produce a sense of common direction or purpose.

There is no one clear and concise definition of the term *strategy* as a management concept. For example, Hofer & Schendel [13] use a resource-based approach and define strategy as the alignment of an organisation's resources with the environmental conditions with which it has to deal. Andrews [14] sees strategy as an outcome of a number of decisions that affect a firm's goals and the stakeholders associated with it. Johnson & Scholes [15] add a scope dimension to strategy by defining it as encompassing all of an organisation's activities and all aspects of the environment. Mintzberg [8] has attempted to reconcile these diverse views by presenting strategy as comprising five different aspects:

(1) Strategy as a plan, where managers map out a conscious intended course of action. The implication is that strategies are made in advance, are deliberate and are developed both consciously and purposefully as a future intent.
(2) Strategy as a ploy, implying a dynamic and competitive element to strategy with manoeuvrings and scheming to the fore, much like a game of chess, where a firm is attempting to outwit its competitors
(3) Strategy as a pattern, seen as a consistency of action and behaviour over time. This perspective views strategy as recognisable because a stream of actions become obvious when looked at historically, as if through a rear view mirror.
(4) Strategy as position, whereby management locate the organisation in its environment. This defines strategy as an outcome of managerial decision making that acts as a mediating force between the organisation and its environment.
(5) Strategy as perspective, whereby the members of an organisation collectively have a deep-rooted perception of the intentions and behaviour of the organisation over time. This definition roots strategy in the collective social understanding of the organisation.

Although Mintzberg has attempted to draw together competing definitions, the fact that there is no one clear view on what strategy is has resulted in a fundamental re-appraisal of the concept by some management theorists. Vasconcellos [16] has returned to the origins of strategy as a concept within military warfare and developed a theory and techniques from this that challenge many of the definitions highlighted above. His contention is that the concept of strategy in modern management thinking has lost its simplicity, value and usefulness. Using the military analogy in business, he differentiates between strategy – *where* to compete and tactics – *how* to compete. Strategy and tactics can have both short- and long-term aspects to them and both can also have degrees of importance attached. Within this perspective there are three types of strategic decisions concerned with *where* to compete:

- In which geographic area does a company wish to compete and/or operate?
- In which industry(s) in a geographical area does a company wish to compete and/or operate?

- In which segment(s) within an industry in a geographical area does a company wish to compete and/or operate?

Vasconcellos sees tactical issues involving the use of functional departments, for example, financial management, accounting, human resource management, production.

Having looked at the concept of strategy, the next section reviews the strategic management process and the activities that managers within a firm go through in order to manage it over the short, medium and long term.

The strategic management process

The strategic management process is the manner in which strategic decision-makers (or strategists) determine the objectives of the firm and make choices to achieve those objectives within the context of the resources available and the firm's mission. The strategic management process has three interlocking parts:

- Strategic formulation
- Strategic choice
- Strategic implementation

Each will be discussed in turn in the following sections.

Strategy formulation

Strategy formulation is concerned with matching the firm's capabilities with its environment and is affected by a number of factors:

- The environmental forces and pressures stemming from the structure of the industry within which the firm competes.
- The resources available to the organisation.
- The internal power relationships within the firm. Organisational politics have an important impact on strategy formulation as established power groupings or coalitions may be threatened.
- The strategists' value systems. Education, family background, attitudes and experiences affect a person's values.
- The firm's history and its relationship to perceptions of the current situation.

Strategy formulation commences with defining (or re-defining) the mission of the firm, its long-term goal. This requires a series of objectives to be set and met with strategies as the means of attaining these objectives. The strategic formulation process requires an assessment of the strengths and weaknesses of the firm (the internal diagnosis) and the opportunities and threats in the environment (the external diagnosis). This assessment is usually termed the SWOT analysis.

Different types of strategy

Strategies form a hierarchy [17]. *Corporate strategy* is concerned with the company as a whole and for large diversified firms it is concerned with balancing a portfolio of businesses, different diversification strategies, the overall structure of the company and the number of markets or segments within which the company competes. *Business strategy* is concerned with a firm's competitiveness in particular markets, industries or products. For a large firm an operational unit will normally be set up, a strategic business unit (SBU) with authority to make its own strategic decisions within corporate guidelines, that will cover a particular product, market, client or geographic area. Finally, the *operating or functional strategy* is at a much more detailed level and focuses on productivity within particular operating functions of the company and their contribution to the corporate whole within an SBU.

The relationship between a firm's mission, its objectives and strategy

The firm's mission is its *raison d'être,* or the fundamental reason why it exists. The firm's mission also sets out what it wants to become over time. Missions can be narrowly or broadly defined and often emanate from the founding entrepreneur's vision of the firm. Businesses do, however, develop, change and diversify but there should be a common thread that links the various parts of the business together. The firm's mission statement should be clearly articulated and allow action to be taken based on it [2]. The mission statement should be:

- Precise
- Indicate how the objectives are to be accomplished
- Indicate the major components of strategy.

Vasconcellos [16] argues that it is important to differentiate between the vision of a strategic business unit and the company as a whole. He proposes that the mission of a business should encapsulate only four elements: the market need being satisfied; the type of clients being served; the geographical area within which the company operates. Again, using a military analogy, Vasconcellos terms these four elements the strategic square, and as components of the mission, they synthesise strategy.

The firm's objectives stem from the mission or mission statement, should be expressed quantifiably and serve four important functions:

- They facilitate performance measurement by allowing actual versus projected performance to be compared and indicate the extent of any gap.

- They can be prioritised, with high order objectives allowing the co-ordination and integration of lower level objectives within the firm.
- They can have a time horizon associated with them which can provide a mechanism for appraising the firm and its managers. If this is to be undertaken then objectives must be specific and verifiable.
- They create a product market focus for the strategy.

A firm's competitive strategy is the distinctive approach the firm takes when positioning itself to make the best use of its capabilities and stand out from its competitors [18], Porter argues that there are four key elements that determine the limits of competitive strategy. These are divided into *internal factors,* first, the company's strengths and weaknesses and second, the personal values of the key implementers and *external factors,* first, the industry opportunities and threats, economic and technical and second, the expectations of society. A firm's strategy is normally defined by four components:

- Its *business scope,* the customers served, their needs and how these are being met. This forms one side of Vasconcellos' strategic square.
- *Resource utilisation,* resourcing properly the areas in which the firm has well developed technical skills or knowledge bases – its distinctive competencies.
- Determining areas of business *synergy,* attempting to maximise areas of interaction within the business such that the effect of the whole is greater than the sum of the parts.
- Determining sources of *competitive advantage.*

Competitive advantage is about being clear where a firm has superiority over competitors, and this may be often located in the technical core of a business. Sources of competitive advantage will be dealt with in more detail in the next sub-section.

Sources of competitive advantage in construction

Determining areas of competitive advantage requires an analysis of the value activities of a company when competing in a particular industry. Value activities stem from the way the company manages its people, the technical system and organisational structure and process, including linkages between inputs from suppliers and the transformation process used by the company to turn these into outputs demanded by buyers. This transformation process, taking inputs from suppliers to create outputs valued by customers, is termed the value chain. It has an external component to the firm, comprising supplier inputs and customer-driven outputs such as distribution channels, each with their own value chains and internal component comprising the transformation process and the manner in which that is supported. The value chain is a product of the firm's history and its strategic management process. The key issue for competitive advantage is the extent to which the firm is able to sustain a long-term advantage through either reducing costs or offering something unique in the way it manages its value chain, which may require reconfiguring.

Porter proposes that competitive advantage can be sustained by being clear about first, the source of competitive advantage within a possible hierarchy of sources. Low order sources of advantage include competing on low labour and materials costs, which are easily replicable by competitors. They form a common basis of competition in construction. Other easily replicable sources include economies of scale stemming from technology, equipment or methods that are also available to competitors. High order sources of advantage include proprietary process technology, which is of particular relevance for competing in engineering process and plant construction, and product differentiation stemming from unique products or services. Other examples of high order sources include brand reputation based on cumulative marketing efforts and customer relationships that are locked to the company through high switching costs, which are also of relevance in many areas of construction. High order sources of advantage require more advanced skills and capabilities to achieve them, often involving specialised and highly trained personnel, or an internal technical capability or close working relationship with leading customers. They also require sustained, cumulative investment leading to the creation of both tangible and/or intangible assets in the form of reputation, customer relationships and specialised knowledge. Second, the number of distinct sources of advantage a company possesses and third, constantly improving and upgrading advantages.

Male [7], using the work of Azzaro *et al.* [19], proposed a value chain analysis of the bidding process within the business strategy system of a contracting company. The value activities in a project for a contracting company are sharply divided between the pre- and post-contract stages. In the pre-contract stage two major value activities proceed in parallel, namely, estimating, and contract planning and management. The estimators analyse a tender using labour and materials constants, seek quotations from subcontractors and utilise learning curves for particular project types to produce a unit rate estimate. Their analysis will not include preliminaries. Contract planners and management, however, are involved in a different process. They adopt a time and resource-based analysis that involves programming site activities, determining the method of working and assessing preliminaries items to produce a 'preliminaries estimate'. As part of this overall process one of the key value activities for the contractor is the sub-contract pricing process and how it fits into the overall pricing strategy. Major decisions include the number of sub-contract work packages and sub-contract quotations sought [20].

The two internal value chain activities are brought together at the adjudication stage of the pre-contract process, where senior management, together with the estimators and contract planners/management involved in preparing the tender, will assess:

- The probable competitors;
- The conditions of contract;
- The client and consultants involved;
- The extent to which the job is required by the company as a project itself and its contribution to workload;
- The estimate of the time likely to execute the project versus that specified in the conditions of contract;
- The relationship between the probability of winning the contract versus the level of mark up and expected profit.

In the estimating and contracts management aspects of the pre-contract phase experience is seen as paramount, and competitive advantage is seen to be gained in the pricing of the preliminaries, where the objective is to devise a programme of work that is shorter than those of competitors. Contractors believe that the preliminaries will win or lose contracts and make a profit or loss. Furthermore, they believe that unit rates in the bill of quantities are unlikely to differ much from competitors. Therefore, in terms of sources of competitive advantage, low order sources are embedded within the estimates produced for the bill of quantities whereas high order sources are embedded in the development of the preliminaries estimate, including the method of working. The next subsection looks at innovation as a source of competitive advantage in construction.

Innovation as a source of competitive advantage in construction

Innovation is one of the key issues in sustained competitive advantage [21]. Clark [22] identifies two types of innovation, radical shifts and gradual incremental innovation. Radical shifts involve short, painful periods of transformation. Most companies experience these very infrequently. Incremental innovation lasts for many years, often goes unnoticed and normally follows radical shifts. In a construction context, Boyd & Wilson [23] concluded that incremental innovation is common in construction whereas radical shifts are rare. Fleming [24] has concluded that product and process innovation in construction is outside the industry's control due primarily to its service nature and the split between production and design, manufacture and construction. He has also proposed that the forms and methods of construction are largely in the hands of the designers.

Male & Stocks [25], drawing together a number of different views on innovation in construction, suggested the following:

- Designers, bring innovation potentially to the process through designing the built form. They provide the spatial and structural form, specifications and drawings that impact the production process through buildability. They largely determine the spatial constraints of the production process for the contractor through the influence of site topography and the planning permissions process. However, the introduction of new materials into the construction industry does not rest purely with designers and is very much dependent on procurement method and contractual arrangements chosen. Materials introduced into the process by designers will only occur directly where nominated suppliers are specified. Where a project is tendered for on the basis of some form of combination of specifications, drawings and bill of quantities the wording of the bill item may allow the contractor to choose the materials or components used provided it conforms to the designer's specification.

- Innovation in materials, as inputs into the production process, does rest outside the contractor's control unless the company has diversified into materials manufacture.
- Innovation in construction plant, forming part of the technical system in the production process, rests with equipment manufacturers and lies outside the industry. However, the choice of plant and the manner in which it is deployed on-site rests with the contractor. The contractor is able, therefore, to draw in plant/equipment innovation into the production process and hence gain potential advantages over competitors. The main issue here, however, is the extent to which high levels of sub-contracting of plant provide competitors with access to the same or similar types of equipment and hence nullify, in a short period of time, any sustainable advantages.
- The primary areas for direct innovation with contractors are:
  - On site level through the 'organisation' of production. Where there is a considerable potential for pre-fabrication off-site the contractor has to consider the timing and scheduling of these inputs together with how they might be fixed within the structure through the correct choice of craft skill. This can again lead to innovation partly in site organisation and partly in craft skills. The level and type of sub-contracting have a significant impact here.
  - At company level by responding to clients with new services and new forms of corporate organisational design.
  - In financial management.
- Codes and standards are ways of achieving technology transfer in construction.

Male & Stocks [25] identified four distinct types of innovation in drawing together their views on competitive advantage in construction:

- Technological innovation, which utilises new knowledge or techniques to provide a product or service at lower cost or higher quality;
- Organisational innovation, which does not require technological advances but involves 'social technology', that is, changing the relationship between behaviours, attitudes and values. Lansley [26] proposed new types of

business organisation, new forms of contract and procurement and the opening up of new markets as examples in construction;
- Product innovation, which may have a low hardware dependency, provides better utilisation of resources and involves advances in technology resulting in superior products or services;
- Process innovations, which substantially increase efficiency without significant advances in technology.

Much of the innovation in the construction industry occurs at the workface, with individual craftsmen, especially on special projects. Contractors that pursue a strategy of a high level of sub-contracting lose this incremental innovation since it is also available to competitors in the marketplace unless long term business relationships exist with sub-contracting firms.

Since innovation in equipment or materials lies outside the industry, construction is more involved with innovation diffusion. However the effective utilisation of equipment and materials by construction firms to maximise the benefits of innovation in other industries creates knowledge-based competitive advantages.

In situations of depressed demand, the emphasis for firms in the industry is on cost reduction and developing new services and competitive advantage for construction firms in the UK has primarily been in organisational innovation. Technology push has also been evident within construction with the development of intelligent buildings and robotics, an issue relevant to Japanese construction firms. In Japan the forces of intense domestic competition, skills shortages and an ageing workforce are a major impetus for the large Japanese construction firms to invest in plant and equipment development to substitute technology for labour. This may well provide them with a competitive weapon to enter overseas markets with product innovations where indigenous firms are not undertaking such development. This is an example of equipment-based innovation in construction in other nations.

In addition, Lansley [26] has identified the importance of understanding the impact of long waves in the economy for the innovatory behaviour of construction firms. Based on his empirical data in UK construction over a twenty-year period, he suggests that different types of innovation may be more appropriate in periods of decline than in periods of recovery. He indicates that in construction during the 1960s companies were involved in process innovation with a primary focus on operational change and a task-oriented style of management. In the 1970s, when construction firms were faced with strategic change, companies were involved in product innovation with a management style that was people and corporate oriented. Flanagan [27], in his analysis of the Japanese construction industry, also provides supporting evidence for Lansley's contention in a cross-cultural context. He indicates that during the period 1977–85 when the Japanese domestic construction industry was in deep recession, a situation of transformational and hence strategic change, construction firms were strategically repositioning themselves by developing new services/product ready for the upturn in the economy. The competitive advantage for Japanese construction firms lies with organisational innovation in the short term and technological innovation for the future.

During periods of competitive change, which required an operational and strategic response, a different type of innovation is required, one that combines both product and process innovation. Process innovation, focusing on gains in efficiency, provides the company with a stable and effective resource base for investment and to launch new competitive initiatives to gain an advantage over competitors.

The introduction of a new wave of procurement routes, such as Private Finance Initiative (PFI) and Prime Contracting also provides opportunities for innovation in financing, design, production and the life expectancy of physical assets. These procurement routes are examples of strategic change since they are fundamentally changing the roles, responsibilities and interactions of the key players in the industry. In addition, procedures such as value management and value engineering also provide opportunities to bring innovation to the process by injecting team-based intellectual capital and innovation at various points throughout project delivery [28]. However, as Standing [29] has argued, there is considerably more potential available for the contractor to bring innovation to the process than is currently the case because procurement routes and the lack of a proactive value management/value engineering incentive programme for the contractor prevent this from happening.

In summary, designers bring potential innovation to the industry through the design of the built form, and through drawings and specifications. Design is a high-order, knowledge based competitive advantage for designers. Consultants are also able to offer new service packages to clients. Innovation for the contracting company is also 'knowledge-based' in that it is concerned with alternative ways of organising the resource transformation process during on-site production, creating new services and designing new forms of corporate organisational structure, or manipulating capital flows. In the UK, technical system innovation lies primarily outside the industry and the contracting company. Skill or craft innovation occurs at the work face and as an asset for the contracting company to use for innovation purposes depending on its strategy towards sub-contracting. Contractors in the UK are 'technological innovators' and not 'technical system innovators' because they determine new ways for the technical systems to interact rather than improve the technical system. However, as mentioned above with the case of the Japanese, this may not always be the case. Innovation in service provision, production and corporate organisation and financial management are 'knowledge-based' innovations and hence are potentially 'high-order' advantages, regardless of whether it is undertaken by consultants or contractors.

This sub-section has reviewed competitive strategy and competitive advantage, the value chain and value activities and the contribution that innovation can make to competitive advantage for designers and contractors.

The next section looks at three techniques that are used in the strategy formulation process, the SWOT analysis, competitor analysis and strategic group analysis, introduced previously in an earlier chapter.

SWOT analysis

The SWOT analysis is shorthand for describing strengths and weaknesses of a company and the opportunities and threats that it faces. Porter [18] contends that there are two forms of strengths and weaknesses facing a company. The first is *structural* and is rooted in the economics of the industry facing the company, coupled with the strategic positioning of a firm relative to its competitors. This type of strength or weakness is external, relatively stable and difficult to overcome. Normally, this will comprise part of the external appraisal of the company to determine the opportunities and threats. Porter describes the second type of strengths and weaknesses as *implementational.* They stem from the company's ability to implement its chosen strategies, the people and management capabilities. Porter sees these as transitory. It is this second set of strengths and weaknesses, the implementational ones, that generally comprise part of the internal appraisal of the company.

The internal appraisal would also normally involve identifying areas of distinctive competence or a competence profile and typical areas covered in the internal audit would be as follows:

- Marketing and distribution channels;
- Production;
- Research and development;
- Human resources;
- Financial resources.

It has been suggested that the primary focus for identifying strengths and weaknesses is in the operating core, that is the input–output transformation process. Glueck & Jauch [2] also argue that the internal audit needs to address two fundamental questions. First, what does the company do well and does this matter? Second, what does the company do poorly and does this matter? These questions must be answered with constant reference to competitors.

Opportunities and threats stem from the external environment and can come from the following:

(1) The economy;
(2) Demographic shifts;
(3) Societal developments;
(4) Technological developments;
(5) Industry structure, suppliers, buyers, competitors, new entrants and finally, the level of competition between companies.

Two strategic management techniques related to understanding the external environment are competitor analysis and strategic group analysis, the subject of the next sub-section.

Competitor analysis and strategic group analysis

Competitor analysis profiles the current and potential future strategies of competitors, it also attempts to work out their possible responses to any changes in strategy the firm may make and to anticipate their reactions to shifts in the task and general environments [3].

Competitor analysis also involves not only understanding existing competitors but also the actions of potential competitors. For each competitor the following factors need to be assayed:

(1) Their future goals at corporate and SBU level;
(2) Their current strategies, both implicit and explicit and their inter-relationship;
(3) The major assumptions held by the company and key managers;
(4) The key capabilities of the company, its distinctive competencies, areas of competitive advantage, growth capability, capacity for the management of change and, finally, the company's staying power in the industry and its chosen strategies.

Pilot research work supervised by Male & Stocks [25] in using competitor analysis for a construction firm indicated that a regional subsidiary of a national contractor operating in both building and civil engineering faced anything up to 55 competitors at national, regional and local levels in all market segments. To make the analysis more manageable, Male and Stocks proposed concentrating on the top 10–15 key competitors, in a firm's peer group to obtain a very good feel for the competition it faces, and the implications that stem from this, in its task environment.

Strategic group analysis, mentioned in an earlier chapter, is attempting to develop groupings of companies in an industry that are following the same or similar strategies. Strategic groups are expected to be relatively stable over time and they affect the pattern of competition within an industry. Where multiple groups exist competition will not be faced equally by all companies. The two most important influences on strategic groups are first, the degree to which the strategic groups interact with each other in the market place and second, the degree of overlap between target customers for each strategic groups overlap. In contracting, procurement and tendering strategies impact and modify strategic groups. Strategic group analysis is explored further in Chapter 5 dealing with the strategic management of consultants and contractors respectively.

Having completed an introduction to the ideas behind the strategy formulation process. The next section explores the ideas behind strategic choice – the decision on which strategy or set of strategies the company should pursue.

Strategic choice

Strategic choice is about making long-term decisions around a range of options that will determine the future strategic position of a firm in terms of its competitors, the strategic group within which it chooses to compete and how it is impacted by the competitive forces that determine industry structure. The strategists should take account of the strengths, weaknesses, opportunities and threats facing the company as well as the constraints that may be impinging on the strategic decision-making process. Strategic choice is also about how the company will seek to attain its objectives in each of the Strategic Business Areas (SBAs) making up its strategic portfolio [30]. This sub-section reviews the nature of decisions and decision-making in companies and then proceeds to describe the nature of SBAs and strategic portfolios.

When choosing among options :

- Strategic decision-makers, the strategists, when making choices, are not totally constrained by environmental, technological or other forces;
- Firms are able to influence their environments and are influenced by them;
- Management perceptions and evaluations mediate between the environment and the actions taken by the organisation to achieve strategic 'fit'.

Strategic choice involves selecting amongst a number of options or strategic alternative(s) that best meet the firm's objectives. The choice will be made using a set of criteria to guide decision-makers. Johnson & Scholes [15] suggest that the eventual decision should be based on the criteria of *suitability* - the degree of fit between the proposed strategy and the situation identified in the SWOT analysis; *feasibility* - the practicalities of the strategy to be adopted; and *acceptability* - once chosen, is the strategy acceptable.

Strategic alternatives

Depending on the analysis of the firm's strengths and weaknesses and the environmental opportunities and threats, the firm's strategists face a number of strategic options or alternatives on which to compete. Porter [18] identified three generic strategies that form the basis of competition, namely:

- *Cost leadership*, where attention is directed to the cost structure of the firm and there is a relentless focus on controlling costs.
- *Differentiation*, where the firm attempts to create the perception that it is different or unique from its competitors. Whilst the firm's cost structure is not ignored it is not the primary weapon of competition.
- *Focus*, where the firm's attention is focused on a buyer group, a particular product line or geographic market. This is often termed a niche strategy, where the firm carves out a competitive arena as a sub-set of a broader market.

Since the latter strategy can also employ cost leadership or differentiation, it has been argued that there are, in practice, only two major generic strategies: cost or differentiation. There are risks associated with each of the generic strategies and individual strategies require different sets of resources and skills for their implementation.

Each of these generic strategies may require a change in direction and there are a number of directions that strategists can take:

- To opt for operational changes only, whilst environmental changes of which the firm is unaware, are under way.
- To consolidate or stabilise. This is a positive decision to remain within existing markets but to make changes in the way the firm operates and also track changes in the environment.
- To retrench. This may be the least frequently used strategy and is sometimes one that may be overlooked for strategic repositioning and/or redefining the business. It is a difficult option for managers to pursue as divestment may imply failure whereas expansion strategies may be perceived as recipes for managerial success.
- To penetrate existing markets; to develop new products and maintain existing markets; to develop new markets whilst maintaining present products.
- To diversify. There are two forms of diversification, related or unrelated. The former, also called concentric diversification, occurs within the broad confines of the industry within which a firm operates. Related diversification within the construction industry can be extensive in that it can include, for example, building and civil engineering contracting, the production of building materials, plant hire firms, onshore construction of oil-rig platforms and property development. These can all be perceived as related to aspects of the construction industry or to construction activity. Unrelated diversification, also called conglomerate diversification, takes the firm outside the industry, markets or products within which it presently operates. Trafalgar House and BICC are examples of conglomerates with interests in construction, that have followed a strategy of unrelated diversification.

There are three means of achieving strategic development:

- Internally, where the firm invests its own capital to set up and operate a new venture. This option is often the primary vehicle of growth.
- Externally through acquisition or merger. This option is often used where speed is of the essence or when a market is growing very slowly or is stagnant. The biggest problem with acquisition is the integration of the acquired with the acquiring firm or in the case of merger the successful integration of two organisational cultures to produce a new culture that represents something other than the dominance of one culture over another.

- A combination strategy which combines elements of internal and external development through contractual agreements. An example of such a strategy in the construction industry is the use of joint ventures.

Finally, there are two additional approaches associated with diversification, which can be linked with internal or external strategic development. The first of these is integration, either forward or backward, to increase the value added. In these instances the firm, usually through acquisition, moves, in the case of forward integration, towards the eventual end purchaser of goods or services. In the case of backward integration the firm moves towards raw materials supply. In a construction firm forward integration would be in the direction of property development and backward integration would be into building materials production. The second approach is internationalisation, in which the firm expands its geographic operating boundaries and moves from operating in a purely domestic business environment to operating in an international one. Internationalisation can be viewed as a form of geographic diversification.

Having identified the best strategy for achieving the firm's objectives, the next stages of the strategic management process are implementation and feedback.

Strategic implementation and feedback

Strategic implementation and feedback occur through the 'organisation' which, as previously mentioned, comprises people who bring to the world of work their beliefs, values, prejudices and biases. Implementation will involve the organisational structure and the setting up of control and feedback systems to ensure that the strategy as implemented is consistent with the strategy as evolved and decided on during the formulation and choice states of our analytical framework. The problem for the implementation process is that an organisational structure is already in place and working. The strategic management process since it is long-term in its orientation, will involve some fundamental shifts in the structure of the organisation. This invariably will draw out resistance from individuals and groups who may feel threatened by these changes [15].

The seven primary mechanisms for strategy implementation are as follows:

(1) Plans and policies at corporate, strategic business unit, operational and functional levels;
(2) A budgetary framework for resource allocation;
(3) Reward systems;
(4) Political systems;
(5) Control and integration systems through organisational structure, the hierarchy, teams and team management, using rules and procedures;
(6) Training and development systems;

(7) Feedback mechanisms. Feedback, allowing any corrective action to be taken, will be provided to the strategic level by measuring actual performance via the systems and procedures that have been set up against the quantitative criteria established for strategic objectives.

To be successful, the strategy implementation process requires three basic questions to be answered systematically to design the appropriate implementation mechanisms. First, who has the responsibility for carrying out the strategies? Second, what must be done in order for implementation to be successful? Third, how will the implementation process work? Unfortunately, the implementation process is the area most often neglected in the strategic management process. The strategic management process, comprising strategy formulation, strategic choice and strategic implementation and feedback will invariably involve the organisation in change, the subject of the next subsection.

The strategic management process and organisational change

The strategic management process is concerned with answering questions: first, what ought the company to be doing? And second, where should the company be going in the future? Three key factors are central to answering these questions:

(1) Strategists should have a future orientation;
(2) Strategists should have the ability to make decisions about the relationship between the company and the business environment that it faces;
(3) Strategists, as managers of the firm, should be able to manage strategic and competitive change.

The strategic management process is concerned with mapping out the direction of the firm in the future and handling strategic and competitive change. Change in an organisation becomes necessary when there are problems, opportunities or threats associated with the following:

- The external environment;
- Diversification strategies necessitating new company structures;
- Technology;
- People.

The process of strategic management adapts a firm to the changes in its environment and Clark [22] has differentiated between change that is recurrent and change that is transformational. Recurrent change creates an organisational memory through the repetition of activities over different time scales. The organisational memory may be appropriately or inappropriately triggered by events in the environment. Transformational change, on the other hand, modifies the recurrent patterns stemming from the organisational memory, either deliberately or unintentionally. Operational change, is an example of recurrent change since the firm normally has a set of well worked out routines to handle it. However, competitive and strategic change should involve the company in transformational change and requires managers to sense the need for change and then exercise choice, having worked out the appropriate way to handle the change. Additionally, the strategic management process, due to its adaptive function between the organisation and the external environment, should involve change that is, hopefully, deliberate.

Managers, when faced by strategic change, can feel stressed and may revert to five possible modes of decision-making, the first four are inappropriate whilst the fifth is appropriate for dealing with change [31]. First, managers may not notice the triggering event and no serious risks or, potentially, no opportunities are perceived and no change results. The status quo remains even though a response is needed. Second, managers may perceive a need for change but respond only in an incremental manner reflecting a small variation from existing patterns. Strategic diagnosis may, however, have determined that there is a need for substantial change. Third, managers perceive serious risks with both new and existing courses of action and believe no solution can be found, they face a Catch 22 situation and avoid taking any action. Fourth, managers have perceived the serious risks associated with current and new courses of action but feel that there is insufficient time to act. Finally, managers have perceived the serious risks from both the current and new courses of action, believe a solution can be found and have the time to undertake the desired courses of action.

Tichy [31] has suggested the following six procedures for assisting with change management:

(1) Managers should identify strategic and non-strategic types of decisions and commit time and resources to assist their decision-making process;
(2) The highly task oriented, short term perspective generated by an operational focus or a crisis can create 'time-traps' that can result in inappropriate responses to change and should be avoided;
(3) There should be a filtering system to target the appropriate response at the right level in the firm;
(4) There should be a buffering system to allow strategists time to focus their attention on strategic decisions and change;
(5) Champions of strategic change should be created whose purpose is to draw the organisation's attention to the change management process that is being implemented;
(6) The time allotted to strategic and non-strategic activities should be monitored.

The next sub-section looks at change management within construction.

The management of change in construction

Lansley *et al.* [32] in their longitudinal study of regional contractors or regional operating units of national contractors highlighted a number of useful principles surrounding the management of change. They also identified some dysfunctional activities. Flexible firms that were successful at change management and able to adapt to the demands of the environment demonstrated the following characteristics:

- Staff perceived senior managers as committed to clearly defined and stated objectives directed at achieving well defined market goals. These strategists had a clear 'vision' of the future direction in which the firm was to go.
- There was a concern for the welfare of staff.
- There was an internal consistency between the management style adopted by senior managers and that preferred by staff and this was reflected in the organisational 'climate'.
- There were high levels of staff morale and job satisfaction.
- There was a history of effective change management.
- Firms had highly effective market sensing mechanisms in place and used their staff with good market knowledge to their most effective potential.
- Effective corporate planning systems had been set up to develop and review the future strategic options available to the firm.

Dysfunctional activities identified by Lansley *et al.* [32] included almost the mirror image of the above:

- There was a lack of concern with human resource management or a corporate perspective.
- Allied to this lack of attention to people there were also higher levels of organisational politics which, in turn, led to low staff morale and lack of organisational effectiveness.
- In terms of adaptive behaviour, firms that understood the need to scan the external environment had implemented the marketing function. In those firms that had not implemented a marketing function, it was mainly due to a lack of understanding how to undertake environmental scanning rather than the need for it. The need for environmental scanning was particularly acute among medium-sized firms.
- Some managers saw corporate planning and budgetary control as synonymous, they are not. The former is concerned with positioning the organisation in its environment and the latter with internal efficiency.

The nature of professional consultancy work in construction, in comparison to that of contracting, may highlight the need for different types of organisational structure and hence responses to the business environment. Empirical evidence of architectural practices indicates that there are three types of organisational structure:

- the *simple hierarchy*, where relationships are vertical rather than lateral;
- *organic*, where structures are relatively unhierarchical but involve a large number of vertical and lateral relationships; and
- *mixed*, where there is a marked organisational tree structure but this is modified in part by lateral relationships.

Private sector practices can be characterised by either organic or mixed organisational structures. The smallest practices tend towards being strongly organic but even the largest practices with some form of tree structure can have overlays of the organic structure. It can be concluded that in the private sector there is a constant emphasis on fluidity of practice and the importance of informal relationships. The overriding need is to meet those changes in the workload primarily induced by the business environment. The informal and not the formal organisational structure appears to be the key in private sector consultancy practices. Since the competitive fee bid system has become the norm for procuring professional services, consultancy firms, like contractors through their estimating function, are having to consider boundary scanning roles such as 'marketing' and 'proposal preparation'. Eventually the counterparts to contractors' estimating and marketing departments will emerge more forcefully within consultancies of different sizes as environmental change compels them to adapt in the medium to long term. Many of the larger consultancy organisations, especially those operating internationally, already have a clearly defined marketing role, indicating the importance of boundary spanning activities for them.

The next section explores the different types of strategic behaviour that are exhibited by firms as an outcome of the strategic management process and the impact that strategists can have, as mediators between the external and internal environments of the firm, in determining that behaviour.

Strategic behaviour

As firms analyse and respond to the environment through strategic formulation, strategic choice and strategic implementation and feedback, it is possible to discern a series of strategic behaviours. A number of different typologies of organisations and their behaviour have been formulated. The typologies that are most useful for describing the strategic behaviour of the variety of organisations in the construction industry and the implications that stem from them for practice are those of Mintzberg [33], Miles & Snow [34] and Ansoff [5]. Other structural types likely to be met in construction will also be discussed and each of the above will be described in some detail and subsequent chapters will elaborate on these further. Before discussing strategic behaviours, however, the first sub-section will commence with the role of the strategist in determining strategic behaviour.

The role of the strategist in determining strategic behaviour

Traditionally, the primary role of the strategist has been seen as being to:

- Monitor, analyse and diagnose the external environment in order to anticipate opportunities and threats;
- Assess the degree of risk associated with any opportunities in the environment;
- Assess the firm's strengths and weaknesses;
- Match the opportunities present in the environment with the firm's strengths whilst minimising the weaknesses against possible threats;
- Develop strategies, decide amongst alternatives and allocate resources to enable selected strategies to be undertaken;
- Monitor results and take corrective action via feedback.

Depending on the size of the firm, the strategist may be an individual or a team. The environmental diagnosis undertaken by the strategist as part of the formulation process can be affected by a number of factors. These include psychological mood, a person's cognitive structure in terms of attitudes to risk and underlying experience, attitude towards change and whether a proactive versus reactive stance is adopted by an individual. Where the diagnosis and formulation process are undertaken by a team, 'group-think' may be present. Underlying the idea of group-think is the notion that group dynamics change perceptions subtly and increase the tendency towards taking riskier decisions than, for example, any one of the group would take as an individual if away from the collective decision situation. In a team or group situation it is advisable to appoint a 'devil's advocate' who will challenge or question group decisions in order to counteract the possibility of group-think.

Mintzberg's typology

Mintzberg has identified five ideal structural types to describe firms. The *simple structure* has direct supervision as its prime co-ordinating mechanism. The key part of the firm is at the strategic level and as its name implies the structure is simple and is uncluttered by rules and regulations. There is little, if any, in the way of functional departments and there is little planning or training but there is considerable personal interaction. The organisational structure can be described as organic. This type of structure characterises small organisations. A small sub-contracting firm, employing a few operatives would be an example in the construction industry. A one or two person architectural, quantity surveying or engineering consultancy would also typify this structure.

There are bureaucratic forms of organisation that standardise processes or skills. The *machine bureaucracy* uses standardisation of work processes as the prime co-ordinating mechanism. This type of structure characterises large firms in stable environments which use highly routine technology. There are many rules and regulations, functional departments and centralised decision-making that follows the chain of command. There is a sharp distinction between line and staff and a well-developed administrative structure. A building component manufacturer would typically have this structure. The *professional bureaucracy* uses standardisation of skills as its prime co-ordinating mechanism. The operating core or production level is the key part of the firm. Specialists staff this type of firm, i.e. professionals, with highly-developed knowledge and skills who have considerable work autonomy. Decision-making is, therefore, decentralised. Standardisation with the professional structural type occurs outside the organisation through educational and professional institutions whereas the machine bureaucracy standardises work flows within the organisation. A typical example of the professional bureaucracy in the construction industry would be a large engineering, quantity surveying or architectural practice – especially one that has diversified internationally. Some of the quantity surveying practices studied by Male [35] and architectural practices by Hillier [36] would typify this structure.

The *adhocracy* has as its prime co-ordinating mechanism mutual adjustment through interpersonal interaction. This structural type is also common in construction and is typified by the project structure. Adhocracies are team based and temporary in nature. They are staffed typically by professionals with a high level of expertise and flexibility for adaptation and problem solving with minimal supervision. Decision-making is decentralised with democratic decision-making common. Influence in an adhocracy is through professional expertise rather than positional authority. Mintzberg differentiates between the operative and administrative adhocracy depending on the nature of the skills brought to the task. The operative adhocracy is concerned with innovation and problem solving directly for the client, often to contract. A key feature of the operating adhocracy is the blurring of the distinction between operational and administrative tasks, with little differentiation of the planning, design and execution of the work. The professional bureaucracy, as a structural type, 'pigeonholes' the solution for the client and uses convergent thinking whilst the operating adhocracy uses divergent thinking to create an innovative solution. The administrative adhocracy is project-based but with the focus on internal performance rather than for an external client. The administrative adhocracy truncates its operating core such that the administrative component is structured as an adhocracy. This is in contrast to the operative adhocracy where the two are blurred. Mintzberg suggests truncation can take place in three ways. First, where there is a requirement to innovate but where the operating core is to remain as a machine bureaucracy which can be established as a separate organisation. Second, where the operating core is done away with totally and is effectively contracted out to other organisations and third, where the operating core becomes totally automated.

The *divisional structure* has as its prime co-ordinating mechanism the standardisation of outputs. The divisional structure creates a series of relatively autonomous smaller firms with functional structures. Groupings tend to be by markets served and the strategic business unit concept (SBU) is applied to each of the relatively autonomous operating units. Divisionalisation is a response to managing the outcome of a diversification strategy. The divisional structure is common in the construction industry especially among the larger contractors. Portfolio analysis, which will be discussed in Chapter 7, is a particularly pertinent technique to apply to this structure.

Miles' and Snow's typology

The typology developed by Miles & Snow reflects the relationship between managers' perceptions of the environment, the internal power and political structure of a firm and the relationship between strategy, structure and process. This typology has been used by Usdiken *et al.* [37] in a study of the Turkish construction industry and also Spiteri [38] in investigating strategic management in architectural and quantity surveying practices.

Miles & Snow [35] have identified four types. *Defenders* are concerned with stability and efficiency. They will concentrate on a narrow range of products or services and opt for a niche strategy through market penetration There is an emphasis on hierarchical control, with centralised decision-making, limited environmental scanning but intensive planing for cost efficiency. Defenders opt for a focus strategy on the basis of cost leadership. *Prospectors* opt for flexibility and the exploitation of new market and product opportunities offered by the environment. There is a stress therefore on environmental scanning. The structure is flexible, stressing informality with few routines and procedures and it will be decentralised. The prospector in its search for new opportunities is, however, likely to be effective but inefficient.

Analysers attempt to combine aspects of the defender and the prospector, seeking to minimise risk while attempting to maximise the opportunity for profit. Analysers move into new markets or products only after market viability is proven. This type of firm is likely to adopt a strategy of imitation. The structure of the analyser will reflect the dual nature of its operations. Parts of the organisation will have high levels of standardisation with the presence of routines and procedures. Other parts of the organisation will be adaptive and have the characteristics of the prospector. *Reactors* are caught between the other three types and represent an organisation with inconsistent and unstable strategic behaviour patterns which is attempting, unsuccessfully, to pursue one of the other three strategies. Reactors respond, therefore, inappropriately to the environment, have a poor performance and do not pursue a particular strategy in an aggressive manner.

Ansoff's typology

The final typology is taken from Ansoff [5]. Ansoff identifies three modes of managed strategic behaviour. The *proactive systematic mode* occurs where change in the environment is incremental and the firm will use extrapolation of historical trends and performance. This typifies long-range planning as opposed to a strategic management approach. Where environmental change is discontinuous and therefore more difficult to predict, there would be a periodic and systematic assessment of the firm's future direction. Thus, the firm would be adopting a strategic management approach. The *proactive adhoc mode* occurs where there is no centrally guided approach to strategic development. The emphasis is on incrementalism and environmental scanning will take place but not in a planned way.

Where change in the environment is incremental, 'bottom up' procedures would be used from the lower levels of the firm. Initiatives would be periodic, and incremental, that is, based very much on current directions with the emphasis on research and development and/or marketing. Where changes in the environment are basically incremental but periods of discontinuity occur the firm will use a trial and error approach in relation to perceived changes in the environment and focus on what are seen as important issues. Strategic thinking and decision-making is implicit. The *reactive mode* has similarities with the reactor type of firm identified by Miles & Snow. The emphasis is on minimisation of strategic changes. Any changes in performance that have been identified by management will be considered to be operative as opposed to strategic. Where changes in the environment are incremental the firm will rely on trial and error approaches triggered by indicators of poor performance Where change in the environment is discontinuous the firm will search for a panic solution when confronted with a crisis.

Other structural types

Finally, this section will be concluded with a brief overview of other structural types that have been highlighted in the strategic management literature and also have a useful analytical role to play in construction. The *sector structure* [9] interposes an additional layer between divisional managers and the corporate centre. Each sector represents common businesses with a clearly defined identity within an industry. This structural type is applicable in large organisations that are adopting growth strategies where there is a danger that the span of control at the centre becomes too great. A company structured sectorally creates 'superdivisions' and in a construction context may, for example, involve creating a 'superdivision' called 'construction' with other, operational divisions called 'management contracting', 'design and build', 'civil engineering' and 'building'. An additional example would be the creation of geographical 'superdivisions' dealing with all aspects of construction activity in each of the sub-continents [25].

The *holding company structure* [15] can take a number of forms. It may resemble a pure investment company with shareholdings in a series of unconnected businesses and where there is little, if any, control exercised to one where a parent company manages a portfolio of virtually autonomous businesses with different percentage stakeholdings in each. Each business unit is likely to retain its own identity and organisational structure. Hunt [11] identifies *the subsidiary company* as a separate structural type and this is often found in construction operating under the holding company umbrella. The subsidiary company has high autonomy from the centre, the primary link being through a financial reporting and a modified management information system. The *matrix structure* is a combination of structures operating concurrently. It comprises functional departments and products or projects, product and geographical divisions and functional and divisional structures. There are two types of matrix structure – the *temporary* matrix as an example of an adhocracy, where the structure is created as required and then disbanded. Temporary matrix structures are common in construction. Stocks [39] in his analysis of project structures in construction, identified the design team as an inter-organisational project structure. Project structures in construction as a temporary inter-organisational matrix. The *permanent* matrix, as its name implies, has a degree of permanency attached to it and creates two sets of permanent managers with separate responsibilities and line-reporting relationships but whose task activities overlap in some manner. The matrix structure, especially the temporary one, tends towards instability, creates ambiguities and, hence, conflict.

The *multinational company or enterprise (MNC/MNE) structure* is found commonly in the form of an international division managing overseas interests. An extension of this structural type is one where geographically based divisions evolve as part of a multinational organisation and each division operates virtually independently by country. Finally, the *global product or integrated structure* is an alternative to the international division. In this instance, the multinational company is split into product divisions with each managed on an international basis. The use of the matrix structure is also common in the MNE. The MNE will be discussed further in the context of international construction.

The *virtual enterprise structure* is one where partnership, collaboration and networking is the glue that binds the 'organisation' together. The virtual enterprise structure has emerged from organisations that wish to manage the linkages between their internal activities and those in the supply and distribution chains. The rapid development of information technology infrastructures has assisted greatly in this. However, as Johnson & Scholes [15] note, the critical issue with the virtual organisation is the extent to which activities can be sub-contracted out and they propose civil engineering as a case in point. Jarillo [40] stated that extreme forms of sub-contracting could result in strategic weaknesses in the long run due to loss of core competencies and an inability to learn and innovate because activities are not undertaken in-house. In their view, this is also linked with the interplay between explicit and tacit knowledge as part of the knowledge creation process within innovation and the fact that virtuality may delimit innovation to within collaborating firms and not across the whole value system. The European eLSEwise research and development project dealing with large-scale engineering investigated the impact of the information technology-led virtual enterprise in construction.

In summary, strategic management literature has identified a number of different typologies that can be used to describe firms. The implications of these for the construction industry will be developed further in Chapter 5.

Conclusion

The chapter commenced with presenting diagnostic models of the firm, differentiating between the strategic, organisational and production levels and the types of managerial decision-making strategies that will be present within the managerial hierarchy. A firm, as an organisation, is goal directed and has number of co-ordinating mechanisms available to it. However, the basic components of organisational structure were identified as structural complexity, formalisation, both explicit and implicit, and centralisation or decentralisation of power. The idea of an organisation as a role system was also introduced, where the formal and informal structures were differentiated, individual differences identified and the impact of the technical system also highlighted.

The components of the strategic management process have been introduced, namely, strategic formulation, strategic choice and strategic implementation. Different levels of strategy have also been identified, namely, corporate, business and operational strategies. The concepts of the firm's mission, its objectives and strategies available to it have been discussed and also the fact that no one correct clear view on strategy available amongst researchers has been identified, although its links with military science have also been raised. Sources of competitive advantage in construction have been identified. These comprise both high and low order factors and the potential for innovation in construction as a source of competitive advantage was also articulated. The role of the new procurement routes for innovation in construction has also been identified and discussed. Techniques used in the strategic management process has been reviewed, namely, the SWOT, competitor and strategic group analyses. The role of strategic alternatives in a firm's ongoing development has been discussed. The circle has been completed by discussing the fact that the 'organisation' is the mechanism by which strategies are implemented and seven primary mechanisms for implementation have been explored. The change process has been evaluated and the characteristics of successful and unsuccessful firms in construction in managing change have been presented. The manner in which strategists in construction manage the strategic management process results in a series of potentially different types of strategic behaviours and a typology of such behaviours has been presented and discussed.

Some applications of these theories are presented in the next chapter.

References

[1] Langford, D. & Male, S. (1991) *Strategic Management in Construction*. Gower, Aldershot.

[2] Glueck, W.F. & Jauch, L.K. (1988) *Business Policy and Strategic Management*, 5th edn. McGraw-Hill, Singapore.

[3] Simon, H.A. (1976) *Administrative Behavior*, 3rd edn. The Free Press, New York.

[4] Feldman, R.C. & Arnold, H. (1983) *Managing Individuals and Group Behaviour in Organisations*. McGraw-Hill, Singapore.

[5] Ansoff, I. (1987) *Corporate Study*, 2nd edn. Penguin, Harmondsworth.

[6] Male, S.P. (1991) Strategic Management in Construction: Conceptual Foundations. In: *Competitive Advantage in Construction*, S.P. Male & R.K. Stocks (eds), pp. 5–44. Butterworth-Heinemann, Oxford.

[7] Male, S.P. (1991) Strategic Management and Competitive Advantage in Construction. In: *Competitive Advantage in Construction*, S.P. Male & R.K. Stocks (eds), pp. 45–104. Butterworth-Heinemann, Oxford.

[8] Mintzberg, H. (1983) *The Nature of Managerial Work*, 2nd edn. Prentice Hall, New York.

[9] Robbins, S. (1987) *Organisation Theory. The Structure and Design of Organisations*. Prentice Hall, Englewood Cliffs, NJ.

[10] Katz, D. & Kahn, R.L. (1978) *The Social Psychology of Organisations*, 2nd edn. John Wiley & Sons, New York.

[11] Hunt, J. (1986) *Managing People at Work*, 2nd edn. McGraw-Hill, Maidenhead.

[12] Chandler, A.D. (1966) *Strategy and Structure*. Anchor Books, New York.

[13] Hofer, C.W. & Schendel, D. (1978) *Strategy Formulation: Analytical Concepts*. West Publishing Company, St Paul, MN.

[14] Andrews, K.R. (1980) *The Concept of Corporate Strategy*. Irwin, Homewood, IL.

[15] Johnson, G. & Scholes, K. (1988) *Exploring Corporate Strategy*, 2nd edn. Prentice Hall, Hemel Hempstead.

[16] Vasconcellos, E. (1999) *The War Lords: Measuring Strategy and Tactics for Competitive Advantage in Business*. Kogan Page, London.

[17] Wheelan, J.D. & Hunger, T.L. (1987) *Strategic Management*, 2nd edn. Addison-Wesley, Reading, MA.

[18] Porter, M.E. (1980) *Competitive Strategy: Techniques for Analysing Industries and Competitors*. Free Press, New York.

[19] Azzaro, O., Hubbard, J., Robertson, D. (1987) *Contractors' Estimating Procedures: an overview*. Occasional paper. RICS, London.

[20] Flanagan, R. & Norman, G. (1989) Pricing Policy. In: *The Management of Construction Firms: Aspects of Theory*, P.M. Hillebrandt & J. Cannon (eds). Macmillan, Basingstoke.

[21] Kay, J. (1993) *Foundations of Corporate Success: How Business Strategies Add Value*. Oxford University Press.

[22] Clark, P. (1989) Social technology and structure. In: *The Management of Firms: Aspects of Theory*, P.M. Hillebrandt & J. Cannon (eds), pp. 75–91. Macmillan, Basingstoke.

[23] Boyd, A., & Wilson, A. (1995) *Technology Transfer in a Construction Background*. Report No. 32. Science Council of Canada, Ottawa.

[24] Fleming, M. (1980) Construction. In: *The Structure of British Industry*, 2nd edn, P. Johnson (ed.). Unwin Hyman, London.

[25] Male, S. & Stocks, R. (1989) Managers and the organisation. In: *The Management of Construction Firms: Aspects of Theory*, P.M. Hillebrandt & J. Cannon (eds), pp. 93–106. Macmillan, Basingstoke.
[26] Lansley, P. (1981) Corporate dislocation: a threat for the 1980s. *Journal of General Management*, Summer, 1–9.
[27] Flanagan, R., Bennet, J. & Norman, G. (1987) *Capital and Counties Report: Japanese Construction Industry*. Centre for Strategic Studies, University of Reading, Reading.
[28] Kelly, J. & Male, S. (1987) *A Study of Value Engineering and Quantity Surveying Practice*. Final Report, Quantity Surveying Division. Royal Institute of Chartered Surveyors, London.
[29] Standing, N. (2000) *Value Engineering and the Contractor*. Unpublished PhD. University of Leeds.
[30] Rice, I. (1989) Strategic Group Analysis within the UK Construction Industry. MSc Thesis. Heriot-Watt University, Edinburgh.
[31] Tichy, N. (1983) *Managing Strategic Change: Technical, Political and Cultural Dynamics*. John Wiley & Sons, New York.
[32] Lansley, P., Quince, T., Lea, E. (1979) *Flexibility and Efficiency in Construction Management*. Final Report. Building Industry Group, Ashridge Management College, Amersham, Bucks.
[33] Mintzberg, H. (1979) *The Structuring of Organisations*. Prentice Hall, Englewood Cliffs, NJ.
[34] Miles, R. & Snow, C. (1978) *Organisational Strategy, Structure and Process*. McGraw-Hill, Tokyo.
[35] Male, S. (1984) *A critical investigation of professionalism in quantity surveying*. Unpublished PhD thesis, Heriot-Watt University, Edinburgh.
[36] Hillier, W. (1979) *The Structure of the Professions Study*. Unpublished report for the Royal Institute of British Architects. Bartlett School of Architecture and Planning, University College, London.
[37] Usdiken, B., Sozen, Z., Enbiyaoglu, H. (1987) *Strategies and Boundaries. Subcontracting in Construction*. CIB W-65. Discussion Paper. November.
[38] Spiteri, J. (1999) *A Critical Analysis of Occupational and Organisational Strategy in UK Architectural and Quantity Surveying Practice*. Unpublished PhD. University of Leeds.
[39] Stocks, R.K. (1984) The Building Team: An Organisation of Organisations. MSc Thesis. Heriot-Watt University, Edinburgh.
[40] Jarillo, J.C. (1993) Strategic Networks: Creating the Borderless Organisation. Butterworth-Heinemann, Chapter 4. Quoted in Johnson, G. & Scholes, K., *Exploring Corporate Strategy: Text and Cases*, 5th edn, pp. 149–199. Prentice Hall, Hemel Hempstead.

6 Strategic behaviour of construction firms

Introduction

The focus of this chapter is to explore principles that can be established from research into the strategic management of construction firms. The chapter heavily draws on the empirical and theoretical work of Lansley *et al.* [1], Male & Stocks [2], Hillebrandt & Cannon [3, 4], Hillebrandt *et al.* [5].

It has been shown that construction markets are defined in terms of an exchange relationship between a buyer and seller, where the former has a need that the latter can satisfy for a mutually agreed price. Markets for contracting firms are defined first in terms of the end product and then geographic location, with the latter especially important for international work (see Chapter 7 on international construction). The 'construction industry', on the other hand, comprises a series of overlapping geographically dispersed markets that are linked socially and economically by firms meeting market needs through the services they provided, the manner in which they manage their businesses and the inputs and outputs that are managed as part of a value chain. The five competitive industry forces identified by Michael Porter set the broad context for strategic decision-making within the competitive arena of construction, and each of the forces will impact markets differently depending on whether the economic structure is construction by contract or construction for speculative purposes.

Industry and market structure impact a firm through its organisation. Senior managers will make choices about where the firm is to compete, perhaps a regional geographic area or they may choose to specialise by project type or by undertaking either civil or building projects or both. Managers' perceptions of how market structures impact the company can lead to different ways of structuring the organisation.

Research conducted among the major construction firms over the past decade has indicated that the structure of firms in the industry had changed significantly [5]. In 1988, 50% of the top thirty-five firms were family owned or controlled. However, in the period 1993–1995, of the eighteen firms from among the group studied by Hillebrandt *et al.* [7], thirteen were PLCs, three were owned by PLCs or non-UK international firms, two were private limited companies that were family-controlled and of the eighteen firms studied, only five were family controlled or owned. However, there has been a major shift away from family ownership or control in the industry towards managerial control and public financing and ownership of firms through the money markets.

Clients' consultant advisers, through their mediating role between client and contractor, can modify the market exchange relationship, and may well dictate the competitive arena for contractors through the choice of procurement route adopted by the client. The client's consultant advisors represent institutionalised distribution channels for contractors services and contractors have to adjust partly to the effects of this mediating role through their own merchant-producer role, that is, the manner in which they buy in and allocate resources to projects. There are major changes occurring in terms of the balance of the traditional roles between the client's consultant advisors and construction firms through the newly emerging procurement routes such as PFI and Prime Contracting, as well as those inherent in the increasing use of design and build and its variants.

Chapter 3 also identified a number of entry and exit barriers in the domestic and international construction industries that structure contracting into an hierarchical industry of companies, where some firms act as subcontractors and others as main contractors. The industry has been described as fragmented but this view has also been challenged earlier, especially where projects are large and/or complex and also, potentially, international. Fragmentation in the industry acts at a number of levels and is rooted in the geographic distribution of production. Construction is a knowledge and information based industry and there is a strong argument to suggest that economists from other industries have failed to understand and learn from its industrial structure derived from location specific, project-based delivery requirements rather than those derived from volume based consumer-oriented product delivery. For example, at the small firm end of the industry, where projects are small, delivery is relatively simple and is skill and trade based, fragmentation is high. However, as project size increases, measured by value, and as complexity increases, either technologically or organisationally or both, knowledge-based rather than skill-based competitive assets become much more important. Fragmentation reduces dramatically such that competition occurs between a small number of firms who operate either nationally or internationally. In construction, fragmentation is handled through forms of business organisation that should seek competitive advantage from knowledge and information based assets rather than through technologically based assets *per se*.

Finally, the State still has a significant and diffused impact on the industry. A number of examples will highlight this subtle influence. First, through the legal framework of industrial relations by introducing legislation that impacts the workplace. One theme that has involved major intervention by the State for construction in this area has been through legislation to reduce or attempt to eradicate tax evasion by the self-employed. Second, depending on the ideological persuasion of the political party in power, Government can alter the balance between employers and employees. Third, the State can facilitate the introduction of initiatives dealing with reforming the industry, such as the

Latham and Egan reports, the Construction Best Practice Programme and M4I. Fourth, Government has been a major capital procurer of construction services. This has changed recently with the introduction of the Private Finance Initiative, where capital expenditure on construction for government facilities has been transferred into revenue expenditure.

Figure 6.1 presents schematically a model of a construction firm that will be used in this chapter.

Core business and core competencies in construction

An empirical analysis conducted by Hillebrandt *et al.* [5] identified core business activity amongst their sample to include:

- Firms who identified it as contracting (N = 10); of which three identified this as purely Building for their firm, six identified this as Building and Civil Engineering and one identified it as Civil and Process Engineering
- Two firms who identified their core as Contracting and Housing
- Two firms who identified their core as Contracting, Housing and Property
- One firm that identified its core as Contracting, Housing and Minerals
- Two firms who identified their core as Contracting and Mining (and other)
- One firm that identified its core purely as housing.

From this, at a more generic level, managers viewed their core business as those activities that possess some or all of the following characteristics:

- Where the firm has had a long standing interest and has built up a considerable expertise
- Generating a fairly substantial turnover
- They are either profitable or expected to be so
- Where reasonable market growth can be expected or where the firm has a captive market
- Where there are low capital requirements.

Kay [6] defines a firm by its contracts and relationships, and added value is created through the success the firm has in putting these contracts and relationships together. Furthermore, it is the quality and distinctiveness of these contracts and relationships to a firm that provide added value to customers. Core business to Kay [6] is those activities where the firm's distinctive capabilities give it a competitive advantage. Distinctive capabilities stem from three primary sources, namely, organisational architecture, reputation, and innovation.

Organisational architecture is easier to sustain than create and is defined by those sets of relational contracts within or around the firm. Relational contracts

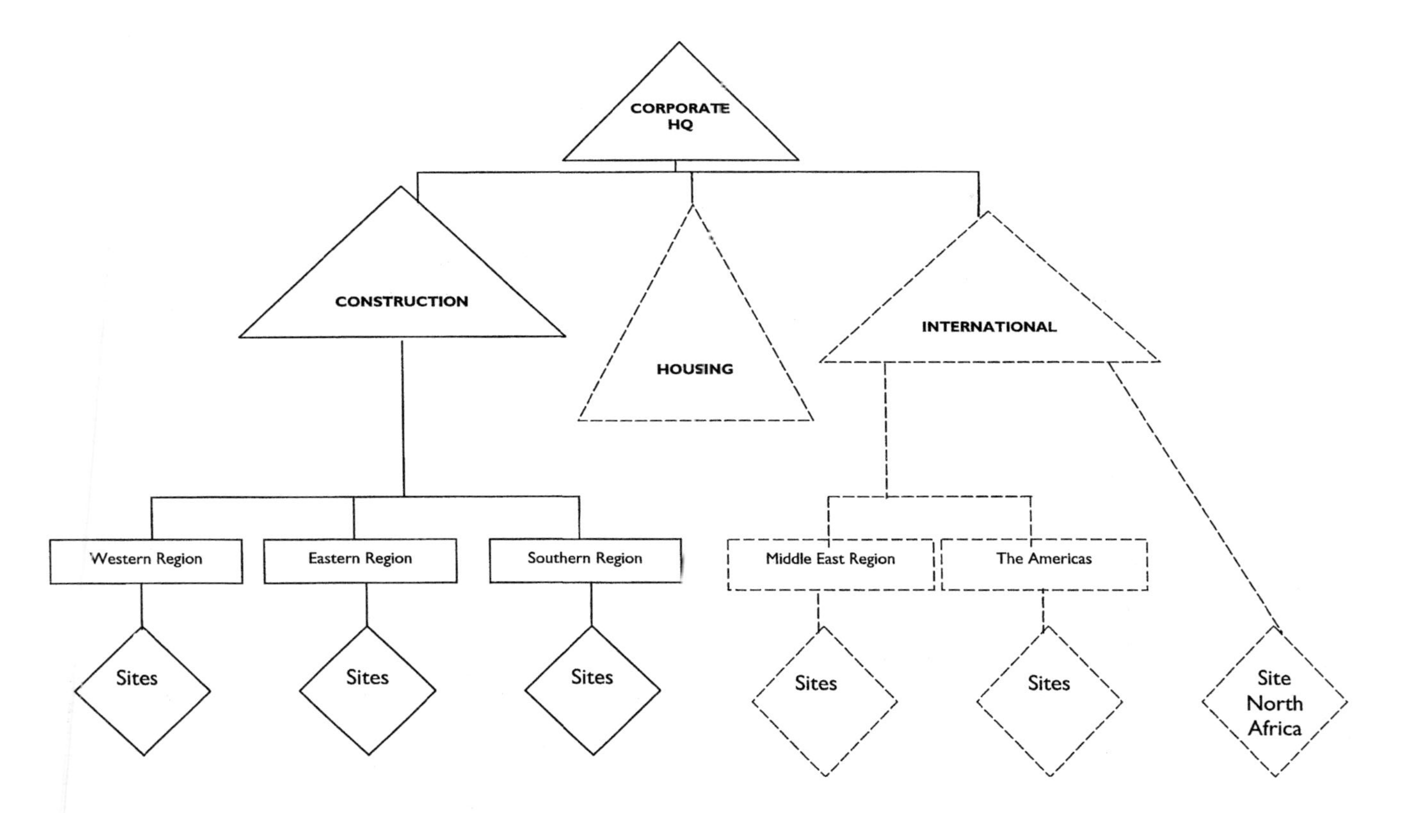

Fig. 6.1 A model of a construction firm (dotted lines indicate possible further organisational developments).

are those that are implicit and are based around expectations and trust between parties. There may or may not be an underpinning legal contract but the relational aspect has both a moral and psychological 'contract' to it that goes beyond any explicit legal contract. Organisational architecture is comprised of three aspects: an internal component, that is, the relationships with and between employees; an external component, that is, with suppliers or customers; or through networks, a group of firms that are involved in related activities. Hence architecture has a social and commercial aspect to it. It allows the firm to create an asset through using these relationships to generate and sustain organisational knowledge and routines that are distinctive to the firm and hence allow it to respond to change and to provide access easily to and for the exchange of information. Organisational knowledge and routines may be generated from the use of technology, or skills or expert knowledge that reside in people's heads. Added value is therefore created from the ability to sustain long term relationships using that knowledge.

The second important aspect of distinctive capabilities is reputation. Reputation conveys information to customers and goes beyond measurable characteristics of any product. For example, most construction firms have pictures of completed projects in their offices or use them as a mechanism to communicate reputation to clients and their advisors. These projects may have been completed within budget, on cost and to the appropriate quality, the usual measurable characteristics of successful project delivery. These pictures present what the firm has achieved. However, pictures alone, whilst they may convey images of impressive projects that have been completed, are incapable of differentiating one firm from others in terms of reputation. They provide no tangible insight or evidence into how the project was delivered by the firm in question compared to other firms. To the client and its advisors it is the manner in which the project was executed to achieve measurable objectives, implicitly embodied in the pictures of completed projects, which is important, perhaps not just on one but on other projects also. Reputation is built out of the continued successful completion of projects and has to be communicated in much more explicit ways than through pictures on walls or in annual reports and marketing brochures. Reputation can also equally be lost and it is easier to lose reputation than build it. The properties of products and services that build reputation can emerge through customer search activity and through comparisons with other similar types of product or service, through immediate consumption (and hence immediate experience) or through long-term experience. The added-value of reputation and the building of it comes more critically from those market exchange relationships where judgements on quality are based on long-term experience as the key customer requirement. The value of reputation as a source of distinctive capability is first, the premium available to be charged due to it compared to the cost of maintaining it. Second, the likelihood of repeat business from a market that is likely to continue once reputation has been built, maintained and communicated in the correct manner. In terms of the characteristics of core business mentioned by Hillebrandt *et al.* [5] it is those markets where a firm has a long standing interest that reputation will provide a distinctive capability.

Innovation is the third source of distinctive capability. Firms may have organisational architectures in place that encourage a continuous process of innovation whilst others may have architectures that are good at implementing innovation. Furthermore, as Kay [6] points out, the process of innovation will often involve complex interactions between firms. However, it is often difficult to create competitive advantage through innovation because it is costly and is also uncertain, it is hard to manage and it is also difficult to secure rewards for the firm and hence make a profit from innovation alone. Therefore, innovation purely on its own may be insufficient to obtain a competitive advantage but when linked with other distinctive capabilities it can become a powerful weapon. Of the two other capabilities organisational architecture is probably the most important when linked to innovation, either through generating innovation, implementing it or both. For contractors much of this occurs at the workface on complex projects, although organisational innovation resulting in offering new services or repackaging existing services primarily appears to occur when recessionary impacts affect a firm.

Core competencies are those competencies that critically buttress the organisation's competitive advantage [7]. They are the collective learning in the organisation, especially in relation to the co-ordination of diverse production skills and integrating multiple streams of technologies [8]. Core competencies differ from organisation to organisation and will depend on its competitive position and the strategies that it is pursuing. There are threshold competencies and resources that are required to remain in the game but these alone will not provide a competitive advantage. Core competencies go beyond threshold competencies, beyond individual business units, and are those that are critical for success in a particular market or industry. They must provide value to the customer, should be difficult for competitors to imitate and are therefore likely to be rare, complex and embedded in organisational knowledge and practice. They are therefore likely to be implicit but need to be made explicit in order to provide a clear understanding of competitive advantage for the firm. Core competencies are not about delivering end-products, for example, the Ford Mondeo in car manufacturing or the office building for a contractor. In these instances the core competency for Ford might be in engine technology or for the contractor the ability to integrate the on-site construction of complex mechanical and electrical installations as well and installing information technology requirements.

Prahalad & Hamel [8] differentiate between core competencies, core products and end-products. In a contracting case, for example, there could be a core competency in constructing highly complex technology-driven designs through onsite production capabilities that underpin the delivery of a range of different end-products, such as office blocks, hospitals and court buildings. However, the core product for the contractor might be offering a capability in design and build or construction management, for example, which is then used to deliver the end-product of the office block for the client. In defining a core competency Prahalad & Hamel [8] propose three acid tests. First, a core competency will provide access to a number of markets such that it can be embedded in different core products. Second, a core competency should also make an important contribution to the perceived benefits that the customer wishes from the end-product. Finally, it should be difficult for competitors to imitate.

In applying the concepts of core business, distinctive capabilities and core competencies, core products and end-products to the construction firm and the data produced by Hillebrandt *et al.* [5], the following picture emerges by way of exploratory example.

(1) Firms may identify contracting as their core business and may realise their competence capability through being the lowest bidder rather than seeking to prevent themselves as having distinctive capabilities or adding value to clients beyond a keen price. The advent of the 'best value' philosophy amongst public sector clients is diminishing the utility of this strategy. The firm that competes on price may have a portfolio of other businesses which, in combination, give them distinctive capability in providing a competitive advantage. An example here would be a firm having a virtually integrated structure such that the businesses complement one another, as shown in Figure 6.2.
(2) A contractor may have a series of long standing relationships, perhaps with key trade contractors, that provide it with a competitive advantage.
(3) For firms that define their core business as contracting, housing and property there are a range of skills that can be unpacked which can be viewed as cross-business, these are core competencies. For example a competency in designing and building high tech buildings within the construction part of the business can be combined with the competencies in housebuilding and property development to enable the firm to deliver PFI packages.

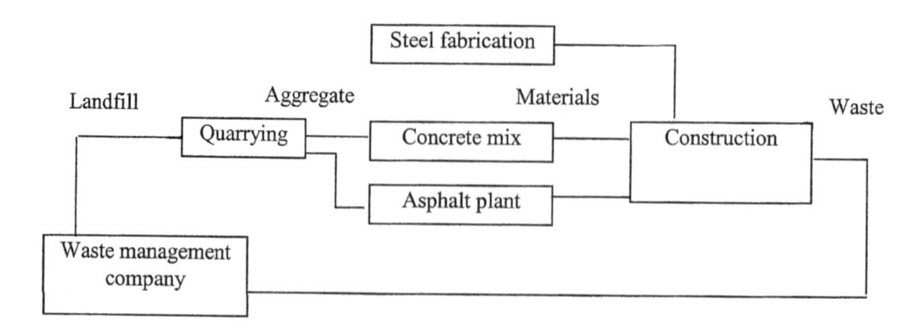

Fig 6.2 Integrated structure of a construction firm. Each of the boxes represents independent firms but by acting together they provide a distinctive capability.

Innovative solutions could include financing, designing and constructing as well as land assembly, land sale and market testing for housing/property investment.

This is using corporate organisational architecture to provide an innovative solution and hopefully build a reputation using this set of competencies to build a core competency. Hillebrandt *et al.* [5] have highlighted in their study the point that during their interviews with senior mangers they found no instance where contractors obtained work due to quality assurance. The possession of BS 5750 is seen as providing the procedural infrastructure to demonstrate that systems are in place. It is a qualifying level of service or threshold level of competency. A focus on delivering quality on projects should be a philosophy that runs much deeper in the organisation within the organisational architecture.

Levels of strategy

Construction firms now have a different set of characteristics from the 1980s and 1990s. This has stemmed from diversification, in its various guises, resulting now in corporate organisational forms for construction firms where the divisional structure is common with a series of separate businesses being gathered under a group or holding company banner. The contracting division will form one part of this wider organisational structure. These new organisational structures have resulted in a requirement to look at strategy and its development at different levels in the construction firm.

At the corporate level of the firm, senior managers will develop a *corporate strategy* that is concerned with balancing a portfolio of businesses. Corporate strategy is company wide and is concerned with creating competitive advantage within each of the businesses a company competes. Corporate strategy is concerned with which businesses should the firm be in and how these should be managed at corporate level. The divisionalised structure, as part of the overall portfolio of businesses, will have different strategic time horizons for each division that has to be incorporated by the main board to produce an integrated corporate strategy. Stokes [9] has suggested there may be a series of possible time horizons occurring within a divisionalised structure. He contends that a corporate level strategic review should take a fifteen-year time horizon. Within that period it is likely that there will be a two-year strategic time horizon appropriate for a contracting division due to the fact that the majority of projects on site do not continue beyond this. A property division would have a four- to five-year time horizon and there would be a very long-term horizon for manufacturing subsidiaries, with detailed analysis not proceeding beyond two years. Other issues that are now present to impact time horizons within divisions will include for those larger firms involved in PFI projects, for example, the fact that they will have 20–30-year time horizons for financing, designing, building, maintaining, operating and potentially transferring back to the public sector. Corporate level strategies will be explored further in the section dealing with strategic choice.

For the larger construction firms, within the contracting division some form of regional 'subsidiary' structure will operate. Regional operating units, or subsidiaries, can be termed *strategic business units* and senior managers at regional level will probably have their own views on how to compete in a particular regional or sub-regional market. Development at this level will require the production of some form of *business unit strategy* that deals with what business the firm should be in at this level and also how it is to compete successfully. Depending on the corporate firm's view on the centralisation or decentralisation of decision-making and the relationship between the region and centre, there will be a choice between top-down decision-making, where decisions are taken at the centre and bottom-up decision-making, where decisions are market led. A strategic choice that involves regional units having decentralised decision-making may require skills at the centre in managing a 'loose–tight' form of organisational structure that combines both a top-down and a bottom-up approach to decision-making. Business unit strategy will be explored in the section dealing with the divisionalised regional structure.

Finally, the operating core or on-site production level involves decisions on an *operating or production strategy*. The competency of site management is a key issue in this respect and they can have a significant impact on production efficiency, especially where subcontractors are concerned. Site management makes many *ad hoc* decisions, often without reference to more senior managers, and they are in a powerful position to adapt and modify the company's policies at production level. With the substantial increase in the use of subcontracting as an operating core strategy the role of the main contractor has moved towards one of managing organisational and contractual risk, the co-ordination and control of boundaries and interfaces between different organisations on site and within the supply chain. Site management is now increasingly required to combine high levels of both technical and managerial competence. Also, the increasing use of sub-contracting has meant that the production process has become fragmented in terms of managerial control, including the intimate and detailed knowledge of production methods and of labour employment. There is a general acceptance of mobility and redundancy by both management and the workforce in construction at operating core level and it is seen as endemic to the structure and operations of the industry. Production strategy will be explored in the section dealing with the operating core.

The next section explores how the strategic management process is and should be undertaken in construction firms.

Managing the diversified construction firm

The strategic management process in diversified construction firms

Construction firms demonstrating good strategic management processes can be characterised by:

- Formulating an overall strategy at the strategic apex that is based on a combination of intuition and informed awareness;
- Expecting operating units to develop and present their own plans to the main board such that they can be consolidated into a single plan;
- Using planning departments to provide contextual background information, undertake analysis and develop the board's thinking into operating plans;
- Having mechanisms in place that permit their strategies to be changed if the external and internal circumstances necessitate it.

In most firms, the planning process is a combination of top-down and bottom-up, that is, the loose–tight approach. The senior managers at main board level provide the goals and vision, whilst those at divisional and regional levels provide the detail and identify the opportunities and actions that are consistent with senior management plans at main board level. The planning horizon is typically three to five years, although the evidence presented by Hillebrandt *et al.* [5] indicates this was getting longer as the effects of the last recession receded. The main area where firms fall down is in the implementation process for plans. First, some firms see strategic planning and the associated plans as the domain of senior management only. This acts against a smooth implementation process. Some firms take a different approach and present plans much more concisely and as a series of discrete business development projects to staff. This assists staff in assimilating the plans throughout the organisation and enables them to focus on well-defined outcomes with those responsible for delivery clearly identified. Regular reviews of progress also assist with the monitoring and implementation process.

Hillebrandt *et al.* [5], as a result of their empirical analysis, put forward a two-phase corporate planning process stemming from the types of environmental change that firms have been and are facing. The planning process comprises two main activities, seeking internal efficiencies and/or responding flexibly to the external environment. The flexibility phase is about formulating questions about the firm's business and identifying alternative paths for its development. The efficient phase is about deciding on a course of action and developing the most efficient way of implementing it. Focusing on short-term budgets is an efficiency issue whereas the use of resources to adapt to changes in the demands of the business environment requires flexible thinking.

During periods when the external business environment is relatively stable, firms do not have to focus their attention on adaptive behaviour *per se* and waste time on flexibility issues. They can focus their attention on preparing detailed plans for achieving high levels of internal efficiency. However, during periods of considerable environmental turbulence, such as those found during periods of deep recession, firms need to attend to flexibility issues and use resources in new ways. During such periods, Hillebrandt *et al.* [5] conclude that focusing on detailed planning for internal efficiency is useless whilst ignoring the requirements to concentrate more importantly on flexibility and the demands of the external environment.

In reviewing management responses to the last recession, Hillebrandt *et al.* [5] conclude there were similarities between the way that firms responded to the recession of the 1970s and that of the early 1990s. The general conclusion is that managers appeared to have learnt little from the events of the 1970s. Firms with more sophisticated corporate planning systems experienced fewer difficulties with coping with the recession than those that had less well-developed systems. Most firms, however, reacted by focusing attention away from the planning process to shorter-term issues and solving immediate financial difficulties and other problems. Even though firms had moved out of the recession of the 1970s and into the 1980s and were thinking about the long-term direction of the firm, when recession hit again in the late 1980s/early 1990s, attention moved inwards and the planning process was seen to be of little benefit. Managers were caught unawares and were 'seriously' surprised by the recession. When the recession took hold, planning was abandoned in favour of short-term answers and a focus on internal efficiency. As the recession receded, more firms began to think strategically, with some becoming committed to strategic planning and some finding it difficult if not irrelevant. Of the firms that had become committed to the planning process, empirical results indicate that only about one third were doing it well. As a final conclusion, the researchers also established that construction firms have developed their own methods of strategic planning rather than follow the planning gurus.

The next section investigates the financial drivers behind the strategy development process in contracting, focusing on how positive cashflows from the construction process can be utilised in assisting strategies of diversification.

Financial drivers behind strategy development

The financial liquidity of contracting, via methods of payment based on regular interim valuations, provides contractors with a number of strategic options that are denied firms operating in continuous process manufacturing. Prior to the Late Payment of Commercial Debt Act (1999) financial liquidity arose from front end loading of contracts, obtaining credit from builders merchants that have payment periods longer than the monthly valuation periods, delaying payment to subcontractors to the end of the month or operating a pay-when-paid policy. Since the advent of the Act such practices have been outlawed and hence the flexibility offered by such arrangements have been limited to the practice of front end loading.

One strategy related to liquidity is to opt to focus on *financial management*, by running down construction operations to reduce overheads, reducing debt, investing cash in the money markets to earn interest or to invest in other types of business. However, if firms chose to opt for the latter, this involves a strategy of *diversification*, which provides contractors with an opportunity to restructure their operations and hence spread risk over different businesses within construction or over different industries. Diversification in construction can occur by the firm taking on work which represents changes in project size, type or location. Additionally, it can include the firm moving into materials manufacture or property development, that is, vertical integration or taking on overseas work which represents geographic diversification and internationalisation. The prompts to internationalise are further discussed in Chapter 7. Empirical evidence indicates that market diversification appears to be the dominant reason for take-over activity in construction and, unlike manufacturing, it is not a strategy to reduce industry capacity.

In the mid 1980s major firms in the industry were keen to borrow money to fuel diversification strategies and invest in property, land for housing, acquisitions in the UK or overseas or both and also to increase their asset base to raise funds for further use. As a consequence of banks' keenness to lend, firms over extended themselves. In 1992 a large number of firms were vulnerable due to heavy debt burdens. The impact on profits was felt during the period 1991–1993, with firms experiencing heavy losses. By 1993 they had been absorbed by the major firms in the industry. In order to address problems in the financial area, firms reduced gearing and their financial difficulties by:

- Selling assets, either those subsidiary firms that required injections of money or profitable businesses, which under normal circumstances, they would have wished to have retained. The consequence of the latter was an imbalancing of their portfolio of businesses.
- Raising capital through rights issues. Approximately 50% of the firms studied by Hillebrandt *et al.* [5] opted for this approach and this appeared to be a strategy adopted by firms in this particular recession.
- Strengthened financial control of the business by improved budgeting procedures, tighter cost controls, improved monitoring of cashflow and better budget planning and use of funds.
- Reducing costs through a number of mechanisms. For example, making redundant permanent staff, re-organising divisions and regional offices to reduce overheads resulting in centralising some activities or conversely reducing head office functions and decentralising services to regions, freezing or cutting salaries and stopping payment of bonuses, reducing the size and number of company cars, cutting back on training and streamlining formal accounting procedures to reduce the number of accounting staff required.

Family firms, perhaps due to their more cautious approach or wish not to dilute family control, appear to have faired better during the recession than other firms studied. Equally, those firms that addressed their financial difficulties early also appear to have coped better with the effects of the recession. However, even though firms took drastic action to bring finances under control, they continued to maintain a policy of paying dividends. This was on the grounds that it kept faith with investors, assisted with further rights issues, supported their share prices and also demonstrated to the Stock Exchange that they were capable of managing the firm financially.

Finally, construction firms in the future will need strong capital asset bases, even for contracting, due to:

- The need to convince clients of their financial strength
- The increasing use of bonding on projects
- The requirement to put equity into certain projects
- The need to raise money on the stock market using a high asset base.

Having reviewed the financial drivers behind strategic choice, the next section will review strategic choice in more detail within the contracting side of corporate configurations.

Strategic change, options and choice in construction

The 1970s witnessed a period of considerable strategic change in the industry and this was repeated in the late 1980s, the early 1990s and is still continuing into the next millennium. Historically, in the 1970s contracting firms were faced with three strategic alternatives:

- To shrink, a retrenchment strategy
- To increase their internal efficiency and exploit their existing markets more intensively – a strategy of expansion within existing markets
- To enter new markets in terms of either project type, size or location, a diversification strategy requiring, in some instances a redefinition of business scope.

In a study published in 1996 Docherty & Langford [10] found that during the recession the firms which employed a strategy of specialisations through divisional strategies were best able to survive the recession. The analysis calculated a Z-score (a multi-attribute measure of financial performance) for ten leading Scottish construction companies and the four most secure companies used divisional strategies. In the period up to the mid 1980s, firms focused their attention on growth and diversification. In the period during the early to mid 1990s firms continued to focus their attention on growth, with firm size continuing to be seen as important and this strategy is still in evidence. However, the last recession has had a significant impact on the way the large construction firms are now managed and structured. The most important impact of the recession on the major construction firms was to force decisions to be made and implemented that were long overdue. Put bluntly, it could be argued that managers during the boom period were to busy focusing their attentions on making money, and consequently, did not see the recession coming and as a consequence senior managers have had to change their management approaches rapidly. The main boards of firms have been reduced in size and become more focused, many firms have also established executive committees below main board level to prepare and read board papers. There has also been a shift in the role of non-executive directors from being one of providing political skills and a network of contacts to providing specialist skills and knowledge to assist senior management teams. Equally, prior to the recession boards did not provide opportunities to develop top class managers and firms were forced to recruit from outside. As a consequence, a more structured management hierarchy has been established with a resulting change of image that has become much more externally focused.

Contractors have also had to explore new ways of obtaining work to maintain turnover related to their core business. They have recognised the need to capitalise on their special expertise and be much more proactive in marketing. This has required them to differentiate their services. They have achieved this through broadening their existing services, offering new ways of packaging existing services, offering new services or taking over unfinished contracts from bankrupt firms in preference to acquiring that business.

In terms of offering different ways of managing projects, design and build has increased significantly in importance as a service, whilst construction management and management contracting have declined in importance – a change potentially in core products. The result on organisational structure has been that divisions that had been set up to handle the latter types of contract have been closed down due to lack of demand and have been incorporated into the normal contracting operations of the firm. In addition, firms have moved downstream in the project life cycle to provide equipment and furnishings fitouts, the maintenance of the structure and also managing the facility. Contractors have also moved upstream in the project life cycle into the pre-construction stage by offering a more integrated design and construction service. They have also become more involved in putting together financial packages and identifying potential projects, of particular importance in overseas markets. Financing has become particularly important in the domestic market with PFI. In addition, some firms have developed specialist expertise in the businesses of their clients after the latter has identified the need. This may have required recruiting specialist staff in order to demonstrate a total capability for delivering the project. Typical areas of specialist expertise developed by firms mentioned by Hillebrandt *et al.* [5] include:

- Water
- Health
- Airports
- Power generation
- Light railways.

There have now been major changes in firms' strategies, summarised as:

- A focus back to core business, as defined by the managers interviewed by Hillebrandt *et al.* [5]

- Increasing attention paid to overseas markets neglected during the boom period prior to the last recession
- Increasing the emphasis on financial management, profits and cashflows
- Increased attention paid to marketing as a business tool
- Tightening up the structure and organisation of firms
- A continued policy of reducing permanent employment that also covers managerial staff
- A shrinking of training and support for education.

Following the recession firms have pursued an objective of constrained growth, partly as a reflection of the over expansion that took place in the late 1980s, with a focus on increased profitability rather than turnover. Firms are also aspiring to remain or become large national or international contractors that need to compete against large mainland European contractors. Some firms are clearly setting their sights on becoming major global competitors. The next sub-sections review some of the options available to contractors and the reasons behind diversification.

Reasons for diversification strategies pursued by construction firms

Five reasons for diversification in construction have been identified

(1) Increasing profitable growth
(2) To seek different activities in which profitable growth could be achieved
(3) Increasing efficiency through control of supplies or link activities because they provided greater synergy
(4) To use positive cashflow and increase fixed assets
(5) To avoid construction cycles and particular clients and markets.

As a result of these reasons for seeking diversification, Hillebrandt *et al.* [5] identified that during the 1980s construction firms diversified into construction-related and non construction-related activities. The researchers argue that their findings indicate that some diversification strategies that had taken place in the 1980s were undertaken for the right reasons whilst others were opportunistic, in areas where contractors had no expertise and as a result were poorly managed. In addition the new acquisitions disturbed the balance of businesses within firms' portfolios. Most firms interviewed as part of their research indicated they would not go into unrelated businesses again. Diversification strategies included moving into:

- Property
- Housing
- Building materials
- Coal and other mining activities
- Plant

- Mechanical and electrical engineering
- Builders' merchants
- Construction businesses internationally.

Non construction-related activities included:

- Timeshare accommodation
- Health care facilities
- Airports
- Waste disposal.

When the recession hit during the late 1980s and early 1990s construction firms, like many firms in other industries, abandoned diversification strategies. However, following the recession, firms have been undertaking diversification again but it is in a different form. It has included moving into areas such as assisting housing associations with skills in the project life cycle (but excluding financing); facilities management; developing complex niche markets that permit the negotiation of contracts, such as in energy projects, airports and health facilities; and also areas that provided income generation capabilities that are independent of the construction cycle.

Centralised versus decentralised strategic decision-making

The issue of centralisation/decentralisation is an important structural attribute of an organisation. In the early to mid 1990s firms viewed centralisation and decentralisation as strategies to achieve economies in staffing and overheads. However, it was not a case of either/or but potentially both operating in tandem for certain organisational functions. A number of firms were keen to run their regional offices as autonomous business units but having strong control from the centre, a 'loose–tight' type of control strategy, as mentioned earlier. Construction firms have also decentralised their market decision-making to divisional boards, which regard themselves as operating near autonomous companies including having responsibility for their own marketing. Following the recession of the late 1980s and early 1990s, all firms centralised financial matters to establish strong controls in this area. There were also cost savings to be made from centralising the accounting function. Also, firms wished to establish a cohesive corporate image, especially when it came to presenting an image of total capability to undertake a wide range of work and to bring all parts of the organisation under one control. However, decentralisation also provides marketing advantages in retaining a local image, it facilitates more rapid decision-making and increased accountability for business units, It also provides and effective means to save overheads. This has been achieved by reducing the number of regional offices operated by a firm but sub-offices may be retained in the larger regions to continue fostering local contacts.

The next section explores the issues of strategy at business unit level within a contracting division, where normally some form of regional structure has been set up.

Business unit strategy: managing divisionalised, regional structures

One of the features of construction that has to be considered in the strategic decision-making process is that any one contract can form a relatively large part of a firm's annual turnover. The decision to commit resources to a particular project can be an important determinant of the profit or loss for any given year. Therefore, the resource capability of the construction firm sets the framework for strategic options such as, for example, growth. Growth for contractors can be achieved in the following ways:

- Through efficiencies only, where no additional resources are required, turnover is maintained, and there is a better use of inputs to achieve efficiency.
- Through growth in size only, a strategy of expansion, where managers focus on the external environment in order to pursue opportunities and less on internal efficiencies.
- Through growth in size and efficiency, requiring managers' attention to be focused externally on opportunities presented in the business environment and internally to increase efficiency.
- Through no growth in size. This is a minimalist strategy and is essentially unstable.

Empirical evidence [11], based on a study of medium-sized contracting firms and regional subsidiaries of national contractors, indicated that a managed growth rate in the region of 10% in real terms per annum is sustainable. As a consequence of pursuing a managed growth strategy, once contracting operations grow to a certain size and geographic spread a regionalised structure needs to be created since it becomes difficult to manage contracts for Head Office continually on a one-off basis. Regionalisation, as a strategy, is the setting up of geographically dispersed operating units or strategic business units (SBUs). The main driver for regionalisation is a growth and expansion strategy. However, a regionalised structure can also involve expansion and retrenchment, sometimes simultaneously. This will depend on how long the regional structure has been set up and the market conditions it faces within different regions or nationally. Retrenchment normally results from declining markets triggered by periods of recession. This may not always be the case, however, since it can also be triggered by other reasons, for example the aggressive actions of competitors and subsequent loss of market share. Regionalisation under both growth and retrenchment will be discussed in this section.

A regionalised structure brings managers in closer contact with the market place, it gains better access to important production and organisational inputs and creates a better working environment for staff at all levels. It can also be a process to decentralise decision-making.

A central theme in creating and managing a regionalised structure within a contracting division is one of managing a process of continuing organisational change and adaptation. Managing a regional structure can involve the setting up or winding down of a series of on-going regional operating units. It also involves decisions on centralisation and decentralisation of decision making within regional structures. Each of these issues will be dealt with in turn.

The processes used by companies to create regionalised units have entailed the following [5]:

- Pilot studies to create organisational units as 'guinea pigs' to develop adequate control and information systems;
- Using specially selected teams well versed in a firm's systems and procedures comprising groups of senior managers, junior staff and former colleagues from headquarters headed by either a prospective or current board member.

These are alternative methods for managing a business project to implement a regionalisation strategy.

Setting up or operating a regional structure also requires sensitivity to the fact that a contracting division is running a series of organisational units that may be at different stages of development with organisational life cycles of birth, growth, maturity and decline. This will require a different set of managerial skills at the centre to handle the different stages of organisational development. For example, the last recession impacted all firms. The larger firms restructured their regional operating units, whilst smaller firms retrenched with the market. Some firms restructured by reducing the number of regional operating units in order to cut overheads but retained sub-offices to retain a local presence. Within this as part of a refocused decentralisation strategy, where estimating, planning, accounting and planning were decentralised, these activities were located at the main regional offices rather than at sub-offices. Also, it was felt important that the regional offices of the large firms needed to provide a presence against competitors in the medium sized firm bracket and senior managers considered that retaining a regional presence was important also for retaining client trust. Equally, it was felt important to retain a regional presence in certain geographic areas because it was essential for winning and executing work. Whilst some firms retrenched or restructured, others, the smaller regional firms, increased their coverage regionally in order to secure work. Depending on the strategies pursued by these firms they could opt for geographic expansion on a project-by-project basis or also consider setting up an embryonic regional structure.

Empirical evidence from the Ashridge studies indicated that 'centres' or head offices had skills to manage regional units in the growth and mature stages but were less skilled at managing regional units in the birth and decline stages. This has been confirmed subsequently in the later studies undertaken by Hillebrandt *et al.* [5] when they reviewed management skills during periods of boom and recession.

In the Ashridge studies conducted in the 1970s that looked specifically at regionalisation issues, nine out of seventeen firms studied that were involved in creating regionalised structures had experienced major problems between regions and the centre. These problems were deeply rooted, had their origins in misunderstandings about the nature of the business by both regional and head office staff and resulted in hostility and mistrust between regions and the centre. For example, one problem highlighted was the centre attempting to apply uniform systems and procedures across all regional units regardless of the size and circumstances of each unit. Smaller regional units were swamped with systems and procedures more appropriate to larger units. As a consequence, smaller regional units were closed down due to poor performance and members of the senior regional management team set up businesses in the locality and became key competitors of the company.

The degree of centralisation–decentralisation from the centre has two further important implications for regionalised structures. The first is the impact on the strategic management process whilst the second is the impact on the training of senior management at regional and head office level. If the centre retains a centralised approach to regional structures, the strategists' focus at the centre is likely to be on the similarity of issues and markets between regions rather than on their differences. However, decentralisation provides regional senior managers with a great deal more decision-making autonomy and allows them to make greater use of local market knowledge. Empirical evidence from the Ashridge studies indicates that many companies failed to capitalise on regional management's market knowledge. Furthermore, those companies that decentralised decision-making avoided centre-region problems. A decision by strategists to define the contracting business from the markets served rather than from the centre requires greater conceptual and judgemental skills. Managers at the centre are managing a diverse portfolio of strategic business units, requiring a mindset able to integrate conceptually what is happening across all regional units.

Centralisation and decentralisation within regional structures also require thinking about senior management succession and associated management development and training. A contracting company defining its business from the centre makes all its strategic decisions at head office. Regional units will be mainly involved in administrative and operational decisions. Senior regional management will primarily be acting in the role of organisational managers and performing an integrative function and using their political and organisational skills [12]. A decentralised regional structure, however, provides a fertile training ground for senior management within the company, both at regional and head office levels. In this instance senior regional managers will be involved in strategic as well as administrative and, to a lesser extent, operational decisions. They are in a better position, therefore, to develop the conceptual and judgemental skills required of the 'institutional manager' [12] at the strategic apex of the company.

Empirical evidence indicates that regionalisation can be successful in companies where the junior staff from head office who are involved in the regionalisation process take over from senior colleagues who return to head office. The second generation of senior regional managers then have experience of both head office and regional procedures and can therefore foster a continued trust from the centre. Regionalisation appears less successful or unsuccessful where a new generation of regional managers are only steeped in regional experience with little or no understanding of the centre. These managers have good knowledge of local conditions but limited knowledge of conditions at the centre. This can lead to an inconsistency of views about the nature of the business between regions and centre. This argues for a rotational management development programme or one where regional managers regularly meet with their head office colleagues.

Strategies at the operating core in contracting firms

As indicated earlier, with such high levels of subcontracting the main contractor's primary role has now become one of organising, co-ordinating and procuring inputs into the production process; providing services of management expertise, experience, backup and resources from an established organisation and an ability to carry contractual risks and obligations for large and complex projects. The on-site production process in construction, unlike the opportunities presented in manufacturing, is characterised by few routine procedures and is labour intensive. The five major inputs into the production process are:

(1) Materials, accounting for between 40 and 50% of production costs [13]. This can vary between 15% for repair and maintenance and up to 60% for contracts with a high building services element.
(2) Labour, which accounts for approximately one-third of production costs [14] but its importance will vary across different types of project.
(3) Site management. Traditionally, site managers on building projects can come from a craft base, are management trainees, or have come through the technical qualifications route or through studying the Chartered Institute of Building examinations and from other building-related disciplines. Site managers in civil engineering have generally come from within the framework of the Institution of Civil Engineers [14].
(4) Plant and equipment, which can be owned, leased or hired. All things being equal, civil engineering projects are normally more equipment oriented than building projects.
(5) Finance for working capital. Requirements for working capital can usually be kept to a minimum due to the monthly valuation process and the

various opportunities available to manipulate finance through credit arrangements with materials and equipment suppliers and payment procedures with subcontractors. The judicious management of capital flows by contractors can result in a positive cashflow and the use of this has been discussed in the section dealing with diversification strategies.

As an important resource into the production process, labour, during periods of boom, can be in very short supply in certain trades. In a recessionary period, operatives and the skilled workforce will be laid off first, with management and support staff retained as long as possible ready for an upturn in the economy. The above resource inputs into the production process will be involved in different intensities during the on-site project life cycle. When aggregated across all sites, the operating core of the contracting firm will be dealing with a diverse range of cyclic inputs across all projects.

The next section reviews ideas behind the consideration of the operating core of the contracting firm as comprising a portfolio of projects with different respective life cycles.

Project portfolios and potential capacity

Each project has a life cycle that goes through the stages of birth (concept and feasibility), growth (design and construction), decay (handover and commissioning) and death (into use). The production level of a contracting company comprises many projects at different stages of their respective life cycles during the construction, handover and commisioning process. The concept of project portfolios is a useful way of thinking about projects within the operating core of a contracting organisation. The usual dimensions for describing a project, and those forming the basis of project portfolios as a concept are:

- Technological complexity;
- Project size, measured by value normally;
- Type of project, for example, office blocks or industrial units;
- Geographic location.

There are conflicting views of how contractors use and perceive a portfolio of projects. Ball [15] takes a market sector view of project portfolios and argues that they provide benefits to the larger rather than smaller contractor by:

- Facilitating the use of differential rates of profit between market sectors;
- Minimising the risk of one contract failing;
- Providing greater bargaining power with clients due to other work in hand;
- Providing gains from acquisition of other contractors in terms of access to non-construction assets, rapid market entry without fear of retaliation,

additional management expertise, portfolios of contracts and contacts and membership of select tendering lists.

On the other hand, Lansley *et al.* [1] propose empirically that contracting companies do not consider demand in terms of market sectors, with the exception of housing development (which has a different economic structure), but in terms of technologies to execute project types. From their perspective, managers assess projects in terms of project size, project complexity, construction method with the associated organisational and managerial requirements of the project.

In combining the various ideas together, the operating core of a contractor will comprise a series of projects at different stages of the their respective life cycle, will vary in terms of size, expressed in value terms, type, complexity, construction method and location. They will also be using a different combination of resources. In traditional manufacturing production capacity is derived from the fact that there is a locationally fixed technical system geared to produce a certain volume of units of output. In contracting, however, the production base is transient, one-off and variable. The operating core of a contracting firm comprises numerous production bases, located in different geographical locations, at different stages in their production cycles and with different resourcing requirements depending on the project life cycle operating at that location. Potential capacity [15] is a useful idea to apply in construction and is the capability of a company to undertake different types of work in the future. It stems from the firm's organisational structure and the accumulated knowledge of management and support functions. Potential capacity is not idle capacity owned or financed by the company but the ability to gear up for a higher workload in the operating core. Where a high level of sub-contracting is undertaken the potential capacity of contracting companies rests from site manager level upwards.

Sub-contracting as a production strategy within project portfolios

Much of the on-site production process is now sub-contracted or directly let by the client out to other firms. This has arisen due to increasing technical complexity of projects, changes in employment legislation over the last twenty years, increasing pressures on employers to reduce fixed costs and the short-term variability of workloads. The result has been a requirement to retain flexibility in the on-site production process. Sub or trade contracting also provides access to specialist knowledge that could be expensive to retain in-house. It is also a low-cost method of organising the work since parts of the production process are sub-let for a known price through competition. Whilst sub-contracting still dominates the operative employment scene there have been strenuous efforts made by the Inland Revenue to get workers back into direct employment.

A number of different types of sub-contractor operate in construction:

- Design, manufacture, supply and fix
- Design, supply and fix
- Supply and fix
- Fix only.

The five most frequently used sub or trade contractors are:

- Heating, ventilating and electrical contractors
- Structural steel and concrete contractors
- Window and curtain walling contractors
- Lift contractors
- Roof, ceiling and floor contractors.

A NEDO study [16] has indicated that there are a number of organisational and contractual problems associated with sub-contracting. Increasing precision is required to define the content and duties for inclusion in sub-contract work packages. The problem is that many things are often left implicit such as sub-contract manning levels, supervision and site programmes. There are often ambiguities surrounding who is responsible for providing plant, tools, power, and assistance with setting out and attendance. NEDO conclude that main contract conditions are no longer an adequate method for regulating the obligations and rights of the main contractor in a situation where much of the work is sub-let to other firms engaged through different contractual terms.

The following summarise the significant contractual issues associated with the employment of sub-contractors [17]:

- A contra charge levied against the sub-contractor for failing to meet the full sub-contract conditions by the main contractor.
- The main contractor holds the sub-contractor responsible for protecting and making good work regardless of when it was installed or damaged.
- The presumption is that the sub-contractor knows the detailed conditions of the main contract.
- Where there is a substantial design aspect by the sub-contractor, the firm is held responsible for design co-ordination with other sub-contractors, especially where there are multiple specialists involved.
- Prior to the Late Payment of Commercial Debt Act 1999 a sub-contractor would often be beset by 'pay when paid' type clauses which protected the cash flow position of the principal contractor to the detriment of the sub or trade contractor.

In order for such conditions to be effective, sub-contracting requires considerable co-operation between site management and sub-contractors but contractual responsibilities are a continual source of divisiveness and also profitability. Sub-contracting also has considerable social costs attached to it, for example, the apportionment of responsibility for health and safety, training and the undermining of the apprenticeship system [18]. This point was also made by Winch [19] who also: argues that sub-contracting puts a break in productivity improvement and that the quality of work done is poorer when sub-contractors are used.

The management resource in construction firms as a source of competitive advantage – resolving a strategic paradox

Strategic choice in construction firms will involve senior managers deciding about a competitive strategy that will enable the firm to exploit its capabilities to the fullest and position itself strategically against its competitors. In periods of boom, the major UK construction companies have often used combined strategies of expansion, retrenchment or stability. In periods of recession, empirical evidence indicates clearly that a common strategy is one of retrenchment, where firms have often being forced into selling business units for the wrong reasons. This has often been coupled with international expansion to offset declining domestic markets although recent empirical evidence has suggested clearly that contractors had lost ground in the international markets prior to the last recession and are now attempting to pull this back.

Hillebrandt *et al.* [5] are of the view that the essential technologies of construction are embodied in the people employed. This is unlike manufacturing businesses where this is vested in plant and machinery. Most plant and equipment in construction is non-project specific, is interchangeable and the important issue is the way that it is used on-site that will determine project success. This is rooted in good selection and planning of equipment for the on-site production process and the use of that technical system compared to potential competitors. Exceptions to the inter-changeability in construction are in areas of special activity such as in tunnelling work.

In order to formulate a competitive strategy it is important to isolate areas where the company has distinctive capabilities that provide it with an edge over its competitors i.e. a competitive advantage. Construction is a knowledge-based industry where similar technologies are available to competitors but it is how they are used that provides the competitive edge. Knowledge resides at regional subsidiary level in terms of detailed market knowledge and client contacts. Knowledge also resides with specialist subcontractors and again this will be available to competitors through the subcontracting process. However, through the use of relational contracts and the use of key supplier lists, contractors are in a better position to retain preferential access to such knowledge. Earlier chapters identified the fact that there is a hierarchy of sources of competitive advantage and that contractors compete on both low and high order factors, some of which are sustainable only in the short term whilst others have the potential to be sustainable in the longer term. Competitive advantage also requires the analysis of value-adding activities and how these are configured into a value chain. Much of the value-adding activities in construction revolve around team and individual knowledge.

Construction relies extensively on project team working and problem solving and isolating competitive advantages in construction can be difficult since there is a high incidence of knowledge-based advantages residing in project teams at different levels in the firm's structure. Some authors have suggested that distinctive capabilities, that provide sources of competitive advantage can be found in the operating core of a company, for example, at site level in contracting. It is often stated that people are one of the key resources in construction and therefore an area of distinctive competence for a contractor could be located at site management level. This would be translated into a distinctive capability that can be termed an operating core *team-specific advantage*. Other areas of distinctive competency that are team-based can also occur in bidding strategy during the adjudication process where key personnel from the middle line and strategic apex come together to produce the final project tender price and also in the selection of plant and equipment and in the planning and scheduling of resources. These are distinctive capabilities that are team-based and reside in the middle line of the firm. These issues are central to the heart of what Chinowsky & Meredith [20] and Chinowsky [21] see as the strategic management process. In a survey of construction companies they found that the issues which drove strategy formulation could be, simplistically, represented by a wheel which depicted seven themes of strategy. This is shown in Figure 6.3.

The value chain has both internal and external components and is a product of a company's history, the impact of the strategic management process and the associated cost and resourcing implications of these. Contractors may be able to get sources of advantage from the way they manage their supply chains as an

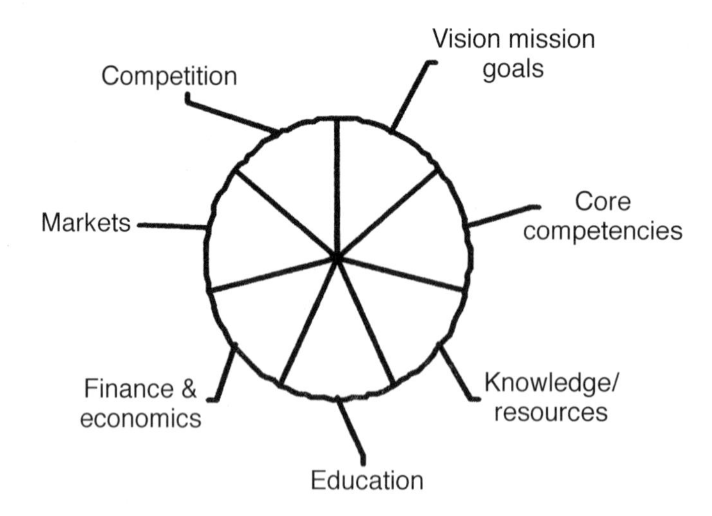

Fig. 6.3 The seven theories of strategy.

input into the production process. It is the quality of the relational contracts that contractors have in place that will be important in leveraging competitive advantage from the supply chain. The management of the supply chain also involves contractors now in offering a range of services to differentiate themselves from the competition. The traditional competitive tendering approach to projects did not allow contractors to easily use their distinctive capabilities to gain a competitive edge since choice of contractor was nearly always dependent on price. However, recent approaches advocated in the Latham and Egan reports to move away from price-based competition alone to one where price and reputation in combination are important for client choice allows differential advantage to come to the fore. Equally, negotiation as a selection process achieves much the same aim.

The preceding is arguing that the management resource at all levels is a key competitive resource. Recent empirical evidence from the Hillebrandt *et al.* [5] research alludes to a different picture in practice, especially when recession hits hard. During the last recession there was a need to undertake substantial cuts in staff numbers. This occurred due to heavy reductions in workload with a consequent need to cut costs. Staffing cuts were both deep and wide across a range of activities and types of employee. The result has been a change in employment patterns. Prior to the recession there was considerable top management mobility and headhunting was prevalent. However, most firms prefer, if possible, to develop managers from within. During the mid 1990s employment of staff tended to be contracted in on a project-by-project basis, with a greater acceptance of agency staff than previously. This is in contrast to the mid 1980s, when there were concerns that this type of employment pattern did not provide staff able to blend in with the company culture. This did not negate, however, the need for a cadre of permanent staff to manage sites but agency and contract staff were seen as less costly and more flexible. The recession of the late 1980s and early 1990s, and for that matter the 1970s, has clearly highlighted a paradox in construction that has implications for competitive advantage. Senior managers have always highlighted the importance of people but the actions taken by senior management when recessionary pressures hit the firm have implied a cavalier treatment of staff. Moreover, whilst the Egan report gives as much attention to education and training as cost reduction and time compression, the training function is seldom given as much attention as factors which are directly, as opposed to indirectly, related to profitability.

Training is a key competitive weapon in a people-intensive, knowledge-based industry. Empirical evidence clearly indicates that graduate recruitment and management training suffered during the last recession. The training function shrank in most firms, with responsibility for training or its costs being devolved to operating units. However, some firms have run counter to this trend and have maintained their training activities and increased the opportunities available to staff. In general firms were unwilling or unable to devote funds to training due to the increased mobility of both middle and senior managers anyway as a prevalent employment pattern. From one perspective, it does seem sensible not to invest in training since it is more likely to benefit the individual's career rather than the firm. However, the counter argument is that without investment in training and management development, knowledge based assets of the firm will not be enhanced and so the firm will lose a source of competitive advantage. Equally, there is a compelling argument to suggest that if construction is about competing on knowledge based assets rather than just price, the time to continue investing in the protection of that asset is during a recession so as to be able to compete more effectively in buoyant times. The turbulence of employment in middle management will have had a deleterious effect upon the perceived attractiveness to potential recruits of the industry. Applicants to University courses in all aspects of the Built Environment (with the exception of Architecture) have declined sharply in the years 1994–1997. Applicants to civil engineering fell from over 5000 to 3600 over the period; Building applicants fell from 4000 to 2700. This fall in student numbers foretells of difficulties in recruiting the next cohort of graduates to take the industry forward [22].

The recession of the late 1980s and early 1990s also highlighted significant problems in the management area. These issues can be grouped under the following headings:

- Diversification strategies.
  - Geographic diversification. Senior management in construction firms took their eye off overseas markets during the preceding boom period, focused their attention on the domestic market and when the recession hit they were unable to regain easily the momentum lost internationally. The overseas market has traditionally been seen as an opportunity to offset falling demand in the domestic market. For some firms this was clearly no longer an option and exacerbated an already difficult situation.
  - Mergers and acquisitions. Managers failed to use their capital wisely, often investing in businesses that were cash hungry, with a result that portfolios became unbalanced. Equally, senior managers had portfolios that comprised many minor business activities. Also, senior management did not appreciate the problems associated with managing acquired firms and how to deal with the issues that arose. This raises questions about the strategic management capabilities of some firms.

- Internal organisation.
  - Within a portfolio of businesses there was strong evidence to indicate that a number of construction divisions were operating without a common policy. Again, this suggests a lack of strategic thinking and awareness.
  - Board meetings that had no adequate control over business units and no agreed actions as outcomes of the meeting. This reflects a lack of ability to provide a sense of common purpose around a clear vision of the firm

and its future. Hillebrandt *et al.* [5] are of the opinion that in the 1990s senior management did not have a clear vision about the future direction of their firms and again this demonstrates a lack of strategic capabilities.

When considering the joint impacts of the above, senior management were demonstrating a lack of strategic awareness over external matters and also a lack of internal management control. This clearly strikes at the need for increased management development in the area of strategic management and again raises questions about why senior management in construction firms appear to set aside the thinking of the strategic management gurus so easily.

There are also two final areas that need to be touched upon when considering the management resource and competitive advantage. First, there is now concern that the UK contracting industry is less able and less well qualified than those managers from some of our mainland European competitor nations, such as France or Germany [23]. This suggests a clear competitive disadvantage internationally, especially given the strategic intent of some of the major construction firms to be able to compete effectively against mainland European competitors and so become global players. Second, Hillebrandt *et al.* [5] found that the role of the personnel function, as an example of support staff in firms, had not improved since the study undertaken in the mid 1980s. The role of most personnel managers was limited and they were seen as administrators of policy made by line managers and not as policy makers. The human resource function appears to be one of the critical roles in continuing to develop a management cadre at all levels within the construction firm with the knowledge, skills and expertise that can be translated subsequently into a source of primary and sustained competitive advantage. This knowledge-based asset at all levels will be located and embedded in the distinctive capabilities stemming from organisational architecture, reputation and innovation.

Conclusions

For the larger contractor, the operating core is in a constant state of geographically dispersed dynamic flux comprised of regionally based and managed project portfolios with different projects at different stages of the project life cycle. The potential capacity to gear up for future workloads resides at regional office and site management levels. The latter is, however, much of the focus for the ebb and flow of contractual difficulties with subcontractors and the translation of company policy into actual site practice. The operating core also comprises different mixes of resource inputs, required at different stages of the project life cycle on-site, different types of subcontractor with different trade union memberships and operative attitudes. These are key inputs from the contractor's supply chain into the production process. Site management is, therefore, a key knowledge-based source of competitive advantage capable of building and losing reputation and also includes the capability to manage and use site-based innovations that occur at the workface. Hence, the site management level is one of the key sources of distinctive capability for competitive advantage in terms of firms' organisational architecture for managing the on-site innovation process and building project site-based reputation for delivery to the client.

The integrity of this organisational architecture is compromised by the long-term subletting of the production process combined with short-term workload variability and the successive impacts of booms and recessions have created a strategic paradox for the contractor. The requirement for organisational flexibility means that the main contractor has to ensure a minimum of capital locked up in fixed and human assets. This creates an unwillingness to invest in management development and training, graduate recruitment and also to subcontracting out much of the production process to others. The operating core of the contracting company is now characterised by a myriad network of sub-contractors of various types, employed with different contractual obligations, rights and liabilities, where, in addition, certain categories of sub-contractors often sublet the fixing and installation of products to self-employed specialists. Main contractors have now become resource acquirers, allocators and managers where traditionally forms of contract have been at the heart of the market exchange relationship at all levels of the firm.

This chapter has highlighted the importance of developing managers at all levels to manage the firm strategically during periods of boom and recession, to manage within operating divisions and within regionally based organisational structures and in the on-site operating core. Managers in the industry have consistently stressed the importance of people in the industry but the evidence suggests that this is not always put into practice. There is an increasing use of agency management staff, project-by-project employment contracts and a view that argues to invest in people is to enhance an individual's career at the expense of the firm's prospects. The previous discussion has proposed that construction is a knowledge-based industry, in much the same way as the information technology software industry or the engineering consulting industry. Knowledge development, maintenance and exploitation is the only sustainable source of competitive advantage for a contractor in the long term, either locally, regionally, nationally or internationally. The empirical evidence presented in this chapter suggests clearly that the strategic actions of managers in the industry pay scant regard to the core competencies that can be developed from knowledge assets in construction.

Hillebrandt *et al.* [5] concluded in 1995 that there is still over capacity in the industry and its structure may well change in the future. They predicted that an elite group of large firms could emerge due to the significant capital requirements for projects and they also foresaw the possibility of the demise of the family firm. The requirement for bonding on projects would also create entry barriers for many firms.

There is no doubt that their predictions about an elite group of UK-based contractors forming in the industry is now well under way. The newer procurement routes of PFI and Prime Contracting will almost certainly cement further an already hierarchical structure in the industry. The newer procurement routes, in particular, coupled with other initiatives under way in the industry will move competition away from a legally based market exchange relationship built around one of competing on price to one of long term relational contracts requiring trust and team working. Provided they can make this transition, the elite group of contractors will be the core constellation of supply chain leaders and global players of the future. Industry restructuring will be addressed further in the chapter on supply chain management.

References

[1] Lansley, P., Quince, T., Lea, E. (1979) *Flexibility and Efficiency in Construction Management*. Final Report. Building Industry Group, Ashridge Management College, Amersham, Bucks.

[2] Male, S. & Stocks, R. (1991) *Competitive Advantage in Construction*. Butterworth-Heinemann, Oxford.

[3] Hillebrandt, P. & Cannon, J. (1989) *Management of Construction Firms. Aspects of Theory*. Macmillan, Basingstoke.

[4] Hillebrandt, P. & Cannon, J. (1994) *The Modern Construction Firm*. Macmillan, Basingstoke.

[5] Hillebrandt, P., Cannon, J., Lansley, P. (1995) *The Construction Industry – In and Out of Recession*. Macmillan, Basingstoke.

[6] Kay, J. (1993) *Foundations of Corporate Success*. Oxford University Press, Oxford.

[7] Johnson, G. & Scholes, K. (1999) *Exploring Corporate Strategy. Text and Cases*. 5th edn. Prentice Hall, Hemel Hempstead.

[8] Prahalad, C.R. & Hamel, G. The core competence of the organisation. In: *The State of Strategy*, pp. 3–15. Harvard Business Review, Cambridge, MA.

[9] Stokes, F. (1977) Practical Problems in Corporate Planning II John Laing. In: *Corporate Strategy and Planning*, B. Taylor & J. Sparkes (eds), pp. 322–333. Heinemann, Oxford.

[10] Docherty, J. & Langford, D.A. (1996) Organisational Structure, Corporate Strategy and Survivability in the Scottish Construction Industry. In: *CIB W65, International Symposium: The Organisation and Management of Construction: Shaping Theory and Practice*. D.A. Langford & A. Retik (eds), pp. 139–154. E. and F. Spon, London.

[11] Building Research Unit (1972) *Efficiency and Growth in the Building Industry*, Ashridge Management College, Amersham, Bucks.

[12] Male, S. & Stocks, K. (1989) Managers and the Organisation. In: *The Management of Construction Firms: Aspects of Theory*. P.M. Hillebrandt & J. Cannon (eds). Macmillan, Basingstoke.

[13] National Economic Development Office (1978) *How Flexible is Construction?* HMSO, London.

[14] Hillebrandt, P.M. (1984) *Economic Theory and Construction Industry*, 2nd edn. Macmillan, Basingstoke.

[15] Ball, M. (1988) *Rebuilding Construction: Economic Change and the British Construction Industry*. Routledge, London.

[16] National Economic Development Office (1988) *Faster Building for Commerce.* Millbank, London.
[17] Building EDC (1984) *Building Skills for Tomorrow's Jobs.* NEDO, London.
[18] Langford, D. (1986) *Labour-only subcontracting.* Technical Information Service paper No. 57. Pp. 1–8. Chartered Institute of Building, Ascot.
[19] Winch, G. (1998) The Growth of Self Employment in British Construction. *J. of Construction Management and Economics.* **16** No. 5: 531–542.
[20] Chinowsky, P. & Meredith, P. (2000) Strategic Management in Construction. *J. of Construction Engineering and Management.* **126** No. 1: 1–9.
[21] Chinowsky, P. (1999) *Strategic Corporate Management for Civil Engineering.* Oxford University Press, New York.
[22] *Building Magazine* 14 July 2000, p. 9.
[23] Clarke, L. & Wall, C. (1998) UK construction skills in the context of European developments. *J. of Construction Management and Economics.* **16** No. 5: 553–567.

7 Strategies for international construction

Background

For many industries, large-scale internationalisation began in earnest after the Second World War, when there was great need for the transfer of technologies, infrastructure and skills to countries that had suffered devastation during the war years. In addition, the ability to communicate, both by data communication and by actual travel, greatly increased as a result of the technology developed during the war and the subsequent reduction in the cost of such communication over the post-war years.

As a result, domestic firms could more easily operate in overseas markets. In a construction context, many of the larger firms began to seek overseas opportunities during the early 1970s due to the decline in demand for domestic construction work. Many firms specifically began to explore opportunities in locations such as the Middle East where industrial, commercial and infrastructure expansion was brought about by a dramatic increase in wealth due to soaring oil prices.

International business strategy

International diversification, or geographic expansion, is not a recent phenomenon with UK construction firms. There was a dramatic increase in overseas work in the 1970s, concentrated mostly in the Middle East. This provided a new source of market opportunity especially in the face of a declining workload in the UK. In practice, a relatively small number of contractors are responsible for much of the UK's overseas export of construction services. Linder [1] recognises that international construction has a history which goes back to Roman times but set off as capitalism internationalised after the First World War.

During the 1980s a decline in oil prices diminished overseas work in the Middle East. However, other regions, such as the Far East, Latin America, Australasia, and Hong Kong, were undertaking large-scale development and infrastructure projects which provided opportunities for UK-based construction and construction consultancy firms. In recent times (the mid to late 1990s) there has been a decline in these markets due to world-wide recession and instability in the financial markets, especially in the Far East. However, many firms still have a significant presence on the world stage and are well established and regarded as international players. Cannon & Hillebrandt [2] have supported this view by suggesting: 'Most of the top 30 companies have at some time operated abroad and most are still doing so.'

Most of these firms saw this as a method of overcoming the effects of the recession in the UK. Construction firms are now differentiating their service by specialising in niche markets, focusing on sophisticated or complex projects, mainly in the infrastructure area or organising financial packages for clients. Contractors, when selecting overseas markets, consider the stability of the country and the ability of to get paid first and then the location decision follows on from that. Most of the major construction firms have established themselves in Eastern Europe, normally on the back of British clients. Some firms have bought interests in former Eastern European contractors in order to increase their presence.

While there may well be a global downturn, international prospects for UK-based construction firms are becoming increasingly important. Table 7.1 shows the balance between new orders for domestic construction and overseas construction (new contracts at current prices). The table shows that the importance of overseas markets in relation to domestic markets has almost trebled in the last ten years or so.

It is crucial that a firm is able to identify strategic opportunities in international markets, and be able to develop competitive strategies to compete successfully in those markets.

Table 7.1 The balance between domestic and overseas new orders (current prices)

	New orders £million	Overseas new orders £million	% of overseas work
1986	17108	1704	10%
1987	22118	2050	9.3%
1988	26298	1387	5.3%
1989	27142	2473	9.1%
1990	22491	2478	11%
1991	19455	2349	12.1%
1992	17493	2991	17.1%
1993	19965	3602	18%
1994	21285	3985	18.7%
1995	22065	5540	25.1%
1996	22834	4812	21%
1997	24806	4062	16.2%
1998	27477	4251	15.5%
1999	26080	3705	14.2%

Note: the percentages have been calculated as a fraction of new orders received in the domestic market.

Source: DETR, Housing and Construction Statistics 1986–2000. [3]

Size and structure

In 1995, the global market for construction, expressed in terms of contracts, awarded to the world's top 225 contractors exceeded $448 000 million [4].

This market is extremely attractive, not only for its size and direct potential for earning profit and generating foreign exchange, but also for the possibility of participation by suppliers of equipment, materials and related construction services.

Most researchers acknowledge that the global construction industry is large. While the ENR report detailed *supra* suggests that the size of the global construction industry in 1995 was around $448 billion, Bauml [5] suggests that this figure may be somewhat overstated due the degree of sub-contracting among the top 225 construction firms.

He goes on to acknowledge, however, that since 'the global construction industry is large, mature, highly fragmented and very competitive' and coupled with the fact that '... no reliable data currently exists for the total global construction industry...' then the size of the industry is '... likely to be far larger than reflected in the ENR survey...'

In general, the increasing demand for construction related work on the global stage may be attributed to a number of factors which include:

- World population growth;
- Higher lifestyle expectations of emerging and third world nations;
- Greater demands for infrastructure and services;
- Growth in aid programmes for agriculture and commerce.

Smith (1991) suggested that many countries are unable to meet the demand for such levels of work, as well as the concomitant financial consequences, and that, as a result, the major construction activity has been concentrated in those countries that have shown a growth in gross national product (GNP). Strassman & Wells [6] concur with this viewpoint by considering that:

> 'As a result of a high level of construction activity over many decades, all of the rich and highly developed industrialised countries have well developed construction industries.'

However:

> 'Many poor countries, on the other hand, with traditionally low levels of construction activity and little industrial and technological development, have not been in a position to exploit and develop their indigenous resources to build up their construction industries.'

They go on to point out that when poor countries '... do acquire finance for modern construction, they find that they lack the skills and know-how, not to mention equipment and materials, for the efficient execution of construction projects...' and that as a result '...such construction resources are imported in the form of a package provided by a foreign or international firm.'

It is against this background that the nature of international or global construction should be viewed. Seymour [7] modifies this background by referring to 'two factors' which have had a significant effect on the international construction environment in recent years. He postulates that the '...first factor to have emerged in international contracting ... is the influx of contractors into the industry from less developed countries...' and the second factor '...has been the lower level of demand world-wide for international contractors services.'

These two factors, together with the historical view of the industry, suggest that, in order to operate successfully in the global construction market, it is necessary to be in possession of a complete, detailed and realistic understanding of the industry and the factors which influence it. As noted previously, the global construction industry is large, mature, highly fragmented and very competitive.

It is vital when examining the global construction market that a detailed appreciation is obtained of where the leading players originate and it has already been pointed out that, in the mid to late 1980s, there was a shift away from highly industrialised countries, as international construction providers, towards less industrialised countries. A detailed understanding of shifts in the size and structure of the market are, therefore, crucial to obtaining a working knowledge of the global construction market.

Table 7.2 contains data on the market share of the top international con-

Table 7.2 Top international contractors market share of awards (%) by country of domicile

Nationality	1991	1993	1995	1999
American	45.7	39.3	16.6	24.1
Canadian	0.8	0.3	0.7	0.0
European	38.3	40.6	50.0	53.6
French	8.1	9.1	15.5	13.2
British	9.4	12.4	4.9	11.7
Italian	7.9	6.8	9.4	2.7
German	7.0	6.5	11.2	10.5
Dutch	1.0	2.4	3.0	3.8
Yugoslav	0.4	0.1	0.0	0.0
Japanese	7.5	13.0	21.3	9.7
Turkish	0.5	0.9	0.0	0.0
Chinese	0.0	0.0	2.8	5.1
Korean	0.0	0.0	4.4	2.3
Other	4.5	3.3	6.0	11.7
All other	7.3	5.9	4.1	5.0
Total	100.0	100.0	100.0	100.0

Source: ENR 2000 [4]

tractors over the period 1990–1999 categorised by the country in which the contractor is domiciled.

It can be seen that the share of awards by US domiciled contractors has decreased significantly from 36.3% in 1990 to 16.6% in 1995, but had recovered to 24% by 1999, while the aggregated European contractors' share has marginally increased from 43.2% in 1990 to 54% in 1999. Compare these figures with the figures for China and Korea for the same period and it can be seen that both these countries showed increases from 0% in 1990 to 2.8% and 4.4% respectively in 1995. Within the European group it can be seen that only the UK has shown a marked drop in share from 10.4% in 1990 to 4.9% in 1995, but again revives by 1999: all other European domiciled contractors (Italians excepted) either retained share or increased share.

The message from these figures is that there would appear to be a shift away from certain industrialised countries as international construction providers, i.e., the US, towards other industrialised countries, e.g. Japan and Germany, as well as towards less developed countries such as China and Korea.

This, of course, is not the complete picture. The data contained in Table 7.1 relate only to international contractors' market share. For a more complete picture of the global industry we must also consider the international consultants' market share. Table 7.3 contains data on the top international designing firms' market share of billings for the period 1990–1995 by country of domicile.

Table 7.3 Top international designers' market share of billings (%) by country of domicile [4]

Nationality	1990	1991	1992	1993	1994	1995
American	42.2	41.1	51.0	42.5	31.5	40.2
Canadian	5.8	6.1	4.4	5.4	6.0	7.4
European	44.9	46.1	38.8	45.7	54.3	42.1
British	17.4	19.3	13.0	13.9	18.9	13.5
German	4.8	5.3	5.5	5.4	6.3	4.8
French	4.8	2.7	4.7	5.0	4.8	3.6
Italian	1.9	1.4	0.6	5.4	1.9	1.7
Dutch	6.7	7.3	7.7	9.0	14.0	11.8
Former Yugoslavia	0.3	0.1	0.0	0.0	0.0	0.0
Other	9.0	10.0	7.1	7.0	8.4	6.6
Japanese	3.2	3.1	2.5	2.6	3.8	4.2
All other	3.9	3.6	3.3	3.7	4.4	6.0
Total	100.0	100.0	100.0	100.0	100.0	100.0

Table 7.3 would suggest that, in general, the market share of billings for international designers appears to remain relatively constant over the period 1990–1995. This is, of course, in direct contrast to the pattern of market share for the top 225 international contractors.

One possible reason for this apparent anomaly is that the shift suggested by Seymour [7], i.e. a move towards contractors from less industrialised countries, is already happening. The question to be posed is: why is this pattern not repeated for international design firms? One suggestion is that the move away from contractors from industrialised countries to those from less industrialised countries is largely due to the use of unskilled local labour by contractors from less industrialised countries thereby creating a lower cost base and resulting in lower tender costs.

The pattern for international designers, i.e. a relatively constant market share of billings over the period, could be explained by the fact the design expertise remains in the possession of the top international designers whereas designers from less industrialised nations have yet to develop such design knowledge and, as a result, have yet to catch up. Tables 7.2 and 7.3, therefore, give an appreciation of the relative market shares of the main providers with regard to the overall size of the international construction market.

An additional factor which is important when considering the market for international construction is the share of work which each market sector generates. Table 7.3 contains a market analysis of the top 225 global contractors by the type of work undertaken in 1995.

Table 7.4 shows that general building with a 37% share is, by far, the largest sector for international contractors. Transportation and petrochemical are the next largest with approximately 18% and 15% shares respectively.

Table 7.4 Market analysis of top 225 global contractors by type of work

Type of Work	1995 Revenue (US$ million)	%
General building	166 200	37.1
Transportation	78 600	17.6
Industrial process and petrochem	65 300	14.6
Power	32 300	7.2
Manufacturing	31 900	7.1
Water supply	18 000	4.0
Sewerage	13 800	3.1
Hazardous waste	6 200	1.4
Other	35 500	7.9
Total	447 800	100

While it is acknowledged that general building work forms the largest single market segment for the top global contractors it must be understood that other factors may influence a contractor's decision to operate in a particular market.

Bauml [5] suggests that '...higher rated E/C [engineering and construction] companies tend to focus their activities on rather larger niches within the industry where they can add value through the application of specialised technical or managerial capabilities.' He highlights some examples of such niche markets including: hydrocarbon processing and power generation. Firms concentrating on niche markets such as these will, undoubtedly, possess specialised technical or managerial capabilities which enable them to position themselves strategically and competitively on the world stage.

While a basic understanding of the size and structure of the global construction market is important it should be noted that it is, essentially, a moving target, i.e. it is constantly shifting and changing. It is, therefore, important that any review of the global market is iterative.

Reasons for internationalisation

Strassman & Wells [6] point out that 'International contracting – or firms from one country building under contract in another – is not a new phenomenon.' As a result of this the reasons why firms choose to operate and compete internationally are relatively well documented.

International contracting may have had its origins with the expansion of empires, e.g. the Romans constructing infrastructure throughout northern Europe, the French constructing the Suez Canal and the Germans constructing the Ankara–Baghdad railway. However, this historical perspective gave way to a new dimension in the 1950s and 1960s when large-scale aid programmes in Third World countries, funded by huge loans from institutions such as the World Bank, the African Development Bank and the Asian Development Bank, were implemented. Following on from this the significant increase in oil exploration in the Middle East, South America and Russia continued to feed the global demand for international construction.

Latterly, the continued growth and development of third world countries and emerging nations, including countries which are the product of the breakup of the former USSR, have given a new lease of life to the international construction market. One-off factors, such as the Gulf and Balkan wars, have also created demand for international construction as part of the rebuilding of infrastructure in these areas.

These factors, therefore, give an indication why firms are compelled to operate on the international stage but the specific reasons are considerably more diverse. The reasons for a firm deciding to internationalise may have their basis in the outcomes of a portfolio analysis. A firm that cannot improve its competitive position within its traditional market can, for example, diversify into other domestic markets, or internationalise.

A portfolio analysis might highlight a number of reasons why a firm should internationalise. These reasons may include:

- The present portfolio no longer meets the firm's objectives. This may be due to:
  - market saturation in the domestic market and unreasonable return on assets;
  - general decline in demand in the domestic market;
  - competitive pressures from other firms in the domestic market.

- The firm may have sufficient cash/resources with which to internationalise. This situation may occur when a firm is meeting its existing objectives in its traditional portfolio and, although further expansion is possible within that portfolio, the firm has sufficient cash reserves to fund overseas expansion.
- Greater profitability is anticipated from internationalisation than diversification. This occurs when the firm has to internationalise because it cannot meet its objectives in its traditional market. Several conditions could cause this to occur:
  - when internationalisation opportunities are sufficiently attractive to offset their inherently lower synergy;
  - when the firm's portfolio contains products/services which are highly sought after in the international arena;
  - when synergy is not considered important by the firm's management and the synergetic advantages of expansion in traditional environments over internationalisation are not important.
- 'Grass is greener' syndrome. This occurs when the domestic market unexpectedly diminishes and the firm has no room to improve its domestic position by competitive analysis. In addition, the firm may have insufficient information to form a complete analysis of the internationalisation opportunities but rather 'dives' into the overseas environment in haste – perhaps because it sees other similar contractors doing the same thing.

The four reasons highlighted above essentially form the basic strategic reasons for a firm's intention to internationalise.

At the outset, the global market is huge. Tucker, R., (personal communication) has stated that 'The engineering and construction industry is global in nature, and the growing markets in the Far East, Middle-East and South America present numerous opportunities for industry participants.' This rosy picture must be tempered by the point made by Maloney & Hancher [8] that 'As the global market place continues to expand ... the competitive pressures on engineering and construction organisations will continue to increase. With the result of reduced profitability'. Why then would firms enter this type of market?

As stated previously, reasons for internationalisation are diverse. Langford & Rowland [9] have identified the following aims for operating in an international constructing environment:

- maximising the company's share value and stock market rating;
- maximising profits;
- maximising dividends;
- securing the viability of the company;
- providing continuity of employment;
- career development and job satisfaction;
- building and maintaining the company's standing;

- maintaining independence and autonomy of the company;
- facilitating corporate growth;
- contributing to the employment and economic development of the international area of operation;
- enhancing or minimising harm to the environment in the area of operation.

Limerick [10] and Imbert [11] have also suggested that some other reasons why firms internationalise may include:

- A desire to increase long-term profitability;
- A desire to maintain shareholders return;
- To spread risk over a greater operating base;
- To balance growth;
- To avoid saturation in established domestic markets;
- To increase turnover.

These reasons were reviewed by Crossthwaite [12] when he carried out research into the reasons why British construction companies internationalised. In his study he found that 70% of respondents felt that a desire to increase long-term profitability was a very important reason for operating overseas. He also found that a desire to maintain shareholders' return was very important for 50% of the respondents and important for the remainder. Balancing growth was very important for 20% of the respondents and important for 40%. The avoidance of saturation in domestic markets was moderately important for 50% of the respondents and important for 40%.

In addition to soliciting contractors' opinions as to the reasons why they desired to operate on an international stage Crossthwaite also sought to establish what the firm's overall objectives were. He found that the firms' main objectives were: to tap new and booming markets; to protect the company against cycles; and, to maintain an edge over competitors.

From the foregoing it can be seen that the main reasons why construction firms internationalise are mainly centred on their desire to increase their profitability, spread their political/economic/commercial risk, avoid saturation in their own domestic market and to improve their competitive position.

However, firms may also desire certain country specific conditions to exist before they would consider entry to a particular country/market. These conditions could include:

- The political stability of the host country;
- The inherent potential economic growth of the host country;
- The existing and potential level of foreign competition in the host country;
- The level of indigenous competition;
- The size of the potential project;
- The historical and potential growth of the market;
- The potential for future projects;

- The links with the home country;
- Market openness;
- Language synergy.

The factors are elaborated on later. In general, there are a range of main reasons why construction firms seek to operate in an international setting. However, it must be noted that individual firms may have their own, very subjective, reasons for seeking to internationalise.

Characteristics and obstacles to internationalisation

Jauch & Glueck [13] have identified that the main characteristics of the international business environment include:

- increased competition
- increased diversity
- increased complexity due to differing:
  - societies
  - cultures
  - educational practices
  - legal frameworks
  - economic and political systems and stability
  - business ideologies.

Additional features may also include:

- inter-governmental relationships
- firm-to-government (both domestic and foreign) relationships, including the need for a sponsor in the target country, e.g. in the Middle East.

In addition to these characteristics of the international business environment, Male [14] has identified a range of characteristics specific to the international construction environment. These characteristics include:

- fragmented industry structure
- geographic dispersion
- decreasing demand with a concomitant 'buyer's market'
- prevalence of 'soft loans', e.g. from ECGD (Export Credit Guarantee Department), in order to secure work
- requirement for host country agent with local knowledge and contacts other risks, including:
  - differing climatic conditions
  - exchange rate fluctuations
  - profit repatriation.

Together these characteristics represent the basic characteristics of the international construction environment, although it must be remembered that specific instances may give rise to additional characteristics which are related to that instance only. Some of the main features you should have considered include: supply chains for labour, materials and money; reputation, both business and technical; capability for providing equity in projects or capacity for the provision of bonds; local representation; understanding of different business and legal cultures; capability to manage geographically distant projects. Notwithstanding the reasons why firms wish to internationalise, consideration must be given to the obstacles and difficulties which may prevent firms achieving their goals.

Foremost amongst these obstacles will be the political risks to which construction organisations are exposed. Wang *et al.* [15] identify several types of risks that occur exclusively in international construction. They list ten risks occurring in Chinese BOT projects which will need to be managed, transferred or accepted. They are risks arising from:

- Changes in the law
- Corruption
- Delay in approvals
- Expropriation
- Reliability and Credit worthiness of client
- *Force majeur*
- Exchange rate and convertibility
- Financial closure
- Despatch and transmission restraint
- Tariff adjustment.

Whilst the study was focused upon China the risks so identified can provide a valuable generic guide for international construction. The issue of free trade has also been picked up by the OECD [16] which has carried out an inventory on the subject of restrictions to trade and found that they included:

- Restrictions on the form of establishment;
- Restrictions on foreign direct investment;
- Requirement of a minimum number of local staff;
- Entry and stay of intracorporate staff subject to labour market test;
- Restrictions on the multi-national relationship of locally established firms;
- Nationality requirements;
- Imposition of a procedure to facilitate entry and stay of professional to supply services;
- Accreditation/licensing of foreign professionals.

In addition, the OECD indicated that restrictions may also exist in areas such as:

- Requirement for local presence;
- Quotas restricting number of foreign professionals/firms to practice;
- Partnership/association joint restriction;
- Restriction on hiring local professionals;
- Government procurement policies;
- Marketing and advertising restrictions;
- Restrictions on levels of fees.

The difficulties highlighted above primarily relate to individuals or firms wishing to operate on an international stage. Langford & Zawdie [17] have argued there may be cultural restrictions which bind companies wishing to operate on an international stage. Different cultures evoke different ways of doing business. Issues such as whether countries are 'deal' focused or 'relationship' focused will shape how negotiations are conducted. Also important will be the level of formality deemed necessary in business relationships and differing attitudes to time can be frustrating for professionals brought up in monochromic business cultures.

Essentially, one of the main difficulties facing a firm wishing to internationalise may be the degree of competition that exists. The degree of competition is usually considerably fiercer than the competition in the domestic market. A cursory glance through international construction journals will reveal that UK firms regularly compete with European, American and Japanese firms for lucrative overseas contracts.

Another difficulty facing a firm will be the degree of financial risk which it will be exposed to. Very frequently the project will be financed by some form of 'soft financing', i.e. ECGD, but in some situations the firm may be required to assist by becoming an equity partner. This can prove problematic when the firm is paid in the local currency, because the firm will then be exposed to the vagaries of currency fluctuations and interest rate differentials.

Geographical dispersion and differing climatic conditions will also prove problematic for the construction firm choosing to operate in the international arena. For example, a civil engineering construction firm based in the UK will be faced with a dramatically different situation if it undertakes work in, say, Malaysia.

In addition, the fact that the firm will be operating under a different economic, political and legal framework will raise issues that it will not face in the domestic market. Many of the opportunities facing UK construction firms in the international arena exist in regions where economic and political stability are well below that which the firm is traditionally used to, e.g. a firm commencing operations in, say, Azerbaijan will be faced with operating in a region with delicate economic stability and an uncertain political future.

Finally, the cultural and business ideology prevalent in the host region may be a complete anathema to the firm wishing to internationalise. This will pose difficulties in setting up the firm's operational base in the host country and will create problems when the firm begins to interact with owners, local regulatory bodies, local suppliers, subcontractors and consultants.

These factors are only representative of the basic difficulties which a firm may face when it wishes to internationalise. Additional factors may arise in relation to specific instances. Despite these difficulties British construction organisations have built up significant investments overseas. At the end of 1998 some £3016m was invested by construction firms. Of this sum over a half (£1636m) was held in North America, approximately a quarter (£688m) in European countries and the balance in Asia and Africa. Taking a snapshot on 1998, inward investment in the UK construction industry by other countries was over £1000m. From these data it would seem that international investments are three times more attractive to UK investors than the UK industry is for overseas investors [3].

Competitive advantage and strategy in international construction

International competition exists in many guises. At one end of the continuum it can take the form of multi-domestic international competition, i.e. it is largely independent within each country. An example of this type of international activity would be banking. Banking will be present in a country that has its own discrete customers, its own assets and its own reputation. At this end of the scale the international construction industry is essentially a collection of domestic industries. Even though there may be multinational firms operating in the industry, each firm's competitive advantages are basically confined to that country or region.

At the other end of the continuum exist global industries, in which a firm's competitive position in one country or region affects its position in other countries or regions. Rival firms compete against each other on a truly global basis and firms attempt to combine advantages in the international arena with those in the domestic arena.

The construction industry in the UK exists somewhere in the middle of this continuum. The vast majority of construction firms operate in the domestic environment because the heterogeneous nature of the industry and the country by country competition makes it very difficult for the majority of UK firms to operate in a foreign country.

However, there are a number of truly internationally focused construction firms in the UK. Cannon & Hillebrandt [2] have suggested that there may be about 30 UK-based construction firms who operate in the international arena. These firms rely on the international market for a significant proportion of their income and they face competition from other international construction firms.

Firms such as Bovis Land Lease, Balfour Beatty and Amec have significant international presence and they compete against other foreign-owned construction firms such as Bechtel, Hyundai Heavy Industries and Kawasaki Heavy Industries for international construction work.

It is important to distinguish between international, multinational and global firms. Figure 7.1 illustrates the differences. While many construction firms can be described as international in scope few would be typified as being 'global' or even multinational. The international firm has a large domestic market and dependent satellites in several countries. The global firm has a home base, but brands independent companies around the world, e.g. Sony (UK).

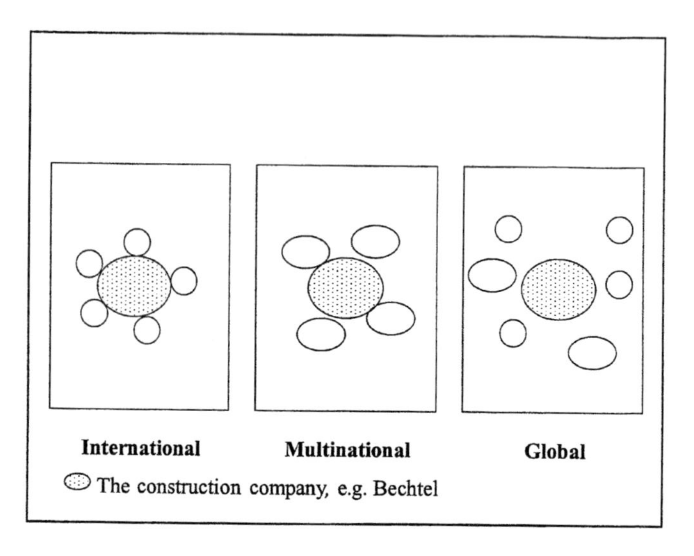

Fig. 7.1 International, multinational and global business (Flanagan [18]).

Competitive advantage in international construction

A major review of the competitiveness of European contractors in a globalising construction market identified Design, Procurement and Construction as core construction industry functions [18]. The contractor's prime function is seen as mobilising the specific resources required for construction of a unique product in a given time. However, as competitive scope is enlarging, the eLSEwise project (referred to in this chapter) also viewed the international contractor's core competencies as widening and defined the emerging international contractor as one that is:

> 'a project centred organisation able to provide flexible logistic skills, manage human resources, provide construction technical skills, organise a network of specialist, have the ability to organise and control financial packages, manage a complex multi-layered and multi-skilled organisation, which in combination can deliver an integrated global offer'. (The eLSEwise project, Male [19].

Hence, the key sources of competitive advantage for international contractors are:

- The ability to provide attractive financial packages. Securing finance is becoming a decisive factor for the viability of many international projects. Providing financial packages has proven to be the most efficient entry barrier for some industry sectors where build, operate and transfer (BOT) and build, operate and own (BOO) projects are rapidly increasing.
- The ability to build winning alliances. The case study based analysis within the eLSEwise project stressed that winning alliances with partners having proven expertise, common experience, compatible business cultures and objectives, derived preferably through a long term partnership approach provides a competitive advantage. Alliances can be global or local, may go beyond the industry boundaries to include clients and financial institutions or they can be short term and project specific or longer term and whole-life focused.
- The ability to accept and manage risk. Clients are increasingly asking contractors to accept a greater share of risk in BOT and BOO projects and this will necessitate risk management throughout the project life-cycle. The implementation of innovation, by its nature, also introduces certain risks into a project and will require technical and managerial expertise to manage those arising from innovation.
- The ability to invest in sales and R&D. The high costs of project development and long tendering and negotiation periods can stretch a contractor's financial capabilities and may act as a strong entry barrier as will investment in R&D activities.
- The identification of client/user needs through market research. Close links with a client preferably developed through long-term partnership relations provide contractors with significant competitive advantage. Changes in the market place, where a contractor has to respond directly to consumer demand stresses the importance of market research as a competitive tool.
- The ability to procure on a global basis. Lump-sum and BOT contracts necessitate searching out cheaper sources of material and equipment and to use lower-cost personal outside of the home country. Computers and related communications technologies make this much easier. The ability to exploit these opportunities better than others and build global networks provides a contractor with a further source of a competitive edge.
- Technical expertise and the right technology. Many international contractors see technical expertise and technology as a source of differentiation from their competitors and they work hard to increase their capabilities in this area. This is the base from which many key success enablers are built, for example, the ability to offer a total skills package, the pre-existence of strong engineering and management skills and the ability for quick mobilisation, the re-use of knowledge, the ability to react fast on any required on-site engineering modification, the ability to offer better engineering solu-

tions, the ability to manage technical risk, the capacity for design management. The question is whether these are threshold competencies or core competencies providing a competitive advantage when unpacked against the competition. Members of the eLSEwise project were clear that technical competency would not be a source of sustainable competitive advantage amongst serious competitors for a project unless it was derived from proprietary knowledge.

- The integration of local and global knowledge. Good local knowledge provides a contractor with an understanding of the socio-economic business environment within which it has to operate. Global knowledge is needed to formulate business and multi-project strategies, to follow market trends and identify potential clients, establish links with potential partners and key suppliers, and to be able to procure globally.
- Political backing. The significance of political backing with political authorities in home and host country has been established as providing a competitive edge. Conversely, close links with authorities in a politically unstable host country can be potentially dangerous.

In the context of understanding the sources of competitive advantage it is useful to understand how these potential sources of advantage can be converted into international success. Here it is useful to apply the model developed by Porter [20] in which he identified four broad attributes of a nation that affect and shape the environment in which domestic firms compete: this environment either promoted or restricted the creation of competitive advantage. These four broad attributes are as follows:

- *Factor conditions* – this attribute deals with the nation's ranking in factors of production required to compete in the construction industry, e.g. factors of production such as skilled labour, skilled management, resources etc.
- *Domestic demand conditions* – this attribute concerns the domestic market's demand for the particular product or service, e.g. construction.
- *Related and supplier/subcontractor industries* – this is concerned with the presence or absence in the domestic market of supplier industries (including subcontractors) and related industries which are internationally competitive.
- *Firm strategy, structure and competitiveness* – this is concerned with the conditions in the domestic market controlling how firms are created, organised and managed. It is also concerned with the nature of domestic competition between firms.

The determinants highlighted above set the scene in which the nation's firms are created. They determine the available resources, skills, goals, pressures, and cultural and behavioural aspects which affect individuals and firms operating in that nation. Firms obtain competitive advantage when the determinants identified above permit the following:

- a rapid accumulation of a set of specialised assets and skills
- the obtaining of better ongoing information and insight into product/service and process requirements
- the goals of managers and employees to support intense commitment and sustained investment
- success in certain industries due to a stimulating and dynamic home environment which encourages firms to capitalise on their opportunities and advantages over a period of time.

Porter's national diamond

Porter [20] suggests: 'Nations are most likely to succeed in industries or industry segments where the national 'diamond' ... a term I will use to refer to the determinants as a system, is the most favourable.' Figure 7.2 is adapted from Porter's 'national diamond'.

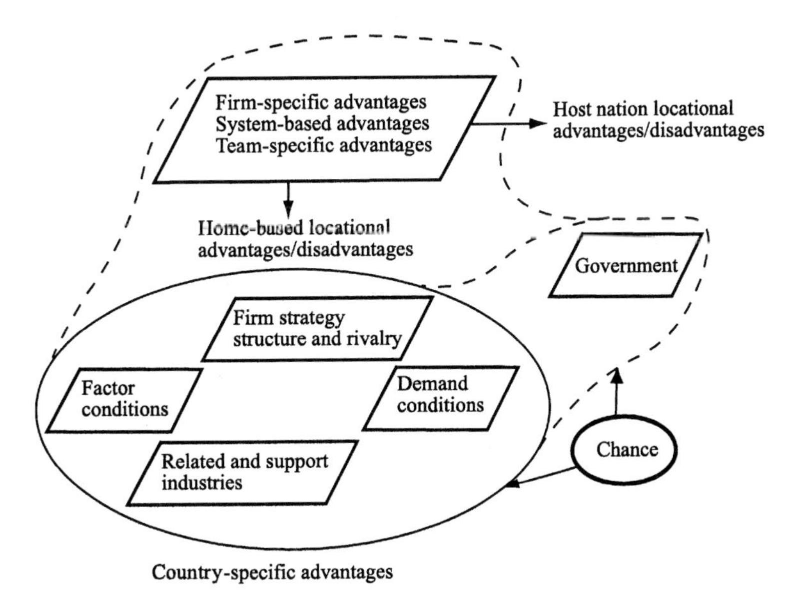

Fig. 7.2 Competitive advantage in international construction. Adapted from Porter [21].

The 'diamond' is considered to be a mutually dependable system, i.e. the effect of one of the determinants is contingent on the state of the other determinants. In addition to the determinants identified above there are a further two important variables which influence the national diamond. These two variables are:

- chance
- government.

Chance

Events brought about by chance cause discontinuities that can permit shifts in competitive position by freezing or re-shaping an industry structure and therefore providing the catalyst for one nation's firms to supplant anothers. Chance events are events which are outside the scope or control of the firm.

Government

Governmental bodies can significantly influence the determinants in the nation's 'diamond'. They can also play a particularly active role in international construction by providing:

- finance
- guarantees (ECGD, etc.)
- promotional support
- diplomatic support.

While the determinants identified above play an important role in setting the scene for internationalisation, the key role is that of the firm itself.

Foreign direct investment

In an international setting it is the economics of the firm that operates globally which is the critical element in the decision to internationalise. Seymour [7] has identified these firms as multinational enterprises (MNEs), which have, as a result of favourable determinants in the home nation (the determinants contained in Porter's 'diamond'), become involved or wish to be become involved in foreign direct investment (FDI). For FDI to take place in international construction, Male [14] postulates that three conditions must be met. These conditions are:

- Ownership advantages – this means that the MNE has competitive advantages over both indigenous firms and other international firms. These advantages can take three main forms:
  - country specific advantages, which applies to both the host country and to the MNE's home country, e.g. sponsorship relationships in the Middle East and joint ventures in Hong Kong
  - industry specific advantages, which may include industries that have particular areas of expertise, e.g. French construction firms who have specialised expertise in the construction of water treatment facilities
  - firm specific advantages, which are derived from the firm itself, e.g. Bovis lend lease with its management contracting leadership.

- Internalised ownership advantages – this means that the firm controls and exploits its ownership advantages itself rather than franchising, licensing or significantly subcontracting to other firms.
- Locational advantage – this condition summarises the principle that it must be more profitable for the MNE to undertake international operations than to undertake domestic operations, i.e. the outcome of the firm's portfolio analysis highlights that greater opportunity lies in internationalisation than in diversification in the domestic market. This condition is measured against the competition from indigenous firms and other international firms.

These three conditions, in addition to the determinants discussed earlier, provide the basis on whether the firm will internationalise or not. It is not a cut and dried decision. The wisdom of internationalising it can only be ascertained by the management of the firm after careful consideration of the opportunities and threats involved.

The MNE/FDI approach, sometimes referred to as the eclectic paradigm, may be more applicable to a manufacturing industry than to the construction industry, and Porter's 'national diamond' model is perhaps more appropriate. However, Male [14] has suggested that when the national diamond is 'placed within the context of the eclectic paradigm, both act together to provide a powerful tool for analysing competitive advantage in international construction'.

As we concluded earlier, an understanding of the determinants that affect the capacity of a firm to internationalise is vital. The determinants contained in Porter's diamond themselves contain key elements that must be identified to enable a complete understanding of the constituent components affecting the conditions for internationalisation.

Factor conditions

These are inputs required by firms in order to compete. They include:

- the aspirations of the board of directors with regard to a global company;
- human resources;
- knowledge of the territory, e.g. geographic, geological and climatic conditions affect a firm to the extent that it becomes adept at competing in an environment which exhibits certain degrees of these conditions;
- knowledge base, e.g. scientific, technical, market and intellectual capital, e.g. the amount and cost of capital available to firms determines how they compete and how effectively they compete;
- infrastructure, e.g. availability of airports, road, rail and telecommunications infrastructure has a greater impact on international construction than it does on domestic construction;
- understanding of the target country's culture, education, law, etc.

Domestic demand conditions

Porter suggests that three broad attributes of domestic demand conditions are significant. They are:

- Domestic demand conditions – these tend to be highly segmented; they are also a key to identifying domestic segments of the market which may be demanded internationally; it may also be extremely demanding clients who require higher standards. This type of domestic client frequently mirrors the type of international client;
- Size of demand – this can lead to economies of scale, increased productivity, increased skills base and increased development in technology. It can have a significant impact on internationalisation where the host country is developing and requires such skills and technology to bring it up to western standards;
- Internationalisation of demand – reduced risks to firm when domestic clients internationalise and decide to use the firm in their international activities.

Related and supplier/sub-contractor industries

Existing synergies with suppliers and sub-contractors may provide international opportunities for firms wishing to internationalise:

- competitive advantage in supplier/sub-contractor industries;
- preferential or advanced access to cost-effective inputs over and above other firms;
- ongoing co-ordination developing into other activities;
- existing synergy providing catalyst for innovation and upgrading examples include: architects who work in the international arena may elect to use home-based contractors, sub-contractors, etc., with whom they have previous relationships;
- development of opportunities for information flow and technical interchange: creation of joint ventures between consultants and contractors who wish to become involved in design and construct work overseas.

Firm strategy, structure and competitiveness

These factors influence the context within which firms:

- compete with each other – the pattern of competition in the domestic market has a profound role to play in the innovation and prospects for international success;

- develop their strategy, both domestic and international - differences in management practices and approaches occur in areas such as: training, background and orientation of leaders as well as the corporate style and the available tools for management. The ability to co-ordinate across functions and the internal attitudes to internationalisation will be an important factor;
- maintain their competitive position - the overall goals of the firm, its management, its employees and its domestic clients all determine how the firm will respond to competition in the domestic market. This will have a material impact on how the firm behaves competitively in the international environment.

These determinants, and their individual elements, provide the basic building blocks upon which country specific advantages can be identified and assessed. In addition, the eclectic paradigm, i.e. the economics of the MNE and the FDI activity, play an important role in modifying the country specific attributes. These and the firm's specific advantages can be pooled together to produce a competitive strategy for a particular firm.

In conclusion, Male [14] has suggested that the following are applicable to competitive advantage in international construction:

- The internationalisation of construction is significantly influenced by early opportunities, advanced and demanding local clients and demand surges in the international arena;
- The competitive advantage in international construction is influenced by existing relationships and synergy between related and support industries and firms;
- Factor conditions - such as available knowledge, education, managerial and operative resources, physical resources, training, and cultural elements - play an important role in providing the correct domestic environment for firms to internationalise;
- Isolation of country specific factors (those contained within Porter's 'national diamond') is vital in determining international competitive advantage;
- Isolation of firm specific advantages, which act as a modifier to the country specific factors, provide the basis by which a firm obtains competitive advantage over other international firms and indigenous firms;
- Isolation of locational advantages permit the firm to make the decision to internationalise into one locus against another. In an international context locational factors primarily relate to issues such as potential market size, political links, inter-governmental relationships, and political and economic stability.

Competitive advantage may not be particularly important in growing domestic markets, but they become increasingly important in declining domestic markets. So far the theoretical aspects of internationalisation have been considered but how do these theoretical aspects relate to the way in which a firm actually identifies, develops and implements an international strategy?

It has been shown that, according to Porter, four broad attributes affect and shape the environment of firms. These attributes include: factor conditions; domestic demand conditions; related and supplier/subcontractor industries; firm's strategy, structure and competitiveness. In addition, Male, has identified that a number of factors influence competitive advantage in international construction.

What steps, therefore, do firms have to take to identify, develop and implement an international strategy to deal with such factors. There is no such thing as a *de facto* generic strategic planning procedure available to the construction firm wishing to internationalise: each firm, and indeed each opportunity, will present a different scenario with its own characteristics, opportunities and threats. Notwithstanding this, a number of key issues will, to a greater or lesser extent, form part of the firm's strategy.

Three main issues are central to the development of a firm's international strategy:

- Country analysis;
- General overview;
- Competitive intelligence.

Country analysis

One of the first issues to be considered when a firm is developing a strategy for internationalisation is the profile of the country in which the firm is seeking to work. It is acknowledged that different countries may hold potential for a firm but it should also be acknowledge that they will also possess their own intrinsic challenges and unique problems. How can a firm assess the viability of entry into one country as compared with entry into another? It is vital that a firm be able to establish a clear picture of the factors applicable to each country in order that a reasoned and unbiased decision can be made.

One of the key starting points for identifying and developing an international strategy is to carry out an assessment of the macro factors which are prevalent in the country or countries being considered. Essentially, a macro overview of a country's economic and political background should be undertaken. This type of exercise is fundamentally a desk exercise using secondary information that permits a macro overview of the country to be obtained.

The key factors that are primarily undertaken in this type of exercise relate to the economic and political profile of the country. The importance of this exercise cannot be underestimated, especially when the firm in question is considering and evaluating long-term potential in the host country.

The first factor to be considered is the economic performance assessment of the country in question. This assessment gauges the country's potential and takes into account the country's past performance, with due regard given to the background and reasons for such performance, as well as extrapolating its future performance. The economic performance indicators which may be used to expedite such an analysis may include: Gross National Product (GNP) in order to obtain a profile of the country's growth over time; details on the level of unemployment, level of wages and educational and training profile of the workforce, in order to ascertain the size and composition of the labour resources available; information relating to the country's level of borrowing from global sources, e.g. the International Monetary Fund (IMF) and the World Bank.

The second factor to be considered by a contractor wishing to operate on the international stage is political risk. Construction by its very nature tends to be an activity which is very often of a considerable duration, e.g. a great deal of construction projects, even of a modest nature, may take upwards of one year while many international infrastructure projects may involve a commitment of several years upwards. This long-term commitment of capital and resources means that any deterioration in the political climate of the host country can mean that the firm's operating position is adversely affected. The effects of such a deterioration could include a number of matters such as:

- the loss of resources, i.e. by enforced take-over with no compensation;
- unreasonable, and enforced, changes with no compensation;
- unreasonable payment retention with little recourse to legal redress;
- the complete cessation of the project with no compensation;
- the detention of staff and possible loss of life.

Ostler [22] has recognised the importance of identifying political risks as part of developing an international construction strategy. He suggests that one of the main systems for identifying political risk is opinion analysis. He suggests that this approach involves the absorption of knowledge from those who are well-versed in the political climate of the country or region in question, e.g. from reading simple political statements to having detailed discussions with people who have an interest in the country or region in question and who are aware of your particular needs and requirements. In addition, use could be made of the appropriate departments within the Foreign Office and the government department dealing with industry as well as regular consultation of the appropriate business and current affairs publications such as *The Economist*.

Ostler goes on to point out that 'Political analysis and prediction can never be one hundred per cent accurate and will always be subjective no matter how experienced and well-versed the source of information is in that particular political climate' [22].

Another approach postulated by Ostler is to build models based on instability measurements. The hypothesis which this approach is based upon is that the greater the political instability then the greater the possibility of change in the political climate. Essentially, the basis of this approach is the taking into account of a number of factors, including:

- the basis of election results
- fall of political figures
- number of assassinations

and weighting them to establish a relatively accurate prediction of political instability.

A final thought on the importance of carrying out a political assessment is suggested by Ostler:

> '...political assessment is important and is a necessary part of the overall analysis, but it will never be an exact science.' [22]

General overview

While a review of the macro issues which are likely to affect a firm's strategy must be undertaken there are a considerable number of other factors which are equally important.

Project size and potential

It is crucial that a review be undertaken of the country's potential as to the type, size and complexity of projects to be undertaken. Essentially, the market for the type of project or projects that are normally carried out by the firm must be established in order for the firm to assess the country's potential over a period of time. This potential must fit in with the firm's capacity and the firm's overall capability and future plans.

Costs and resources

Ostler suggests that these factors should not only be examined at a particular point in time but that consideration should be given to how they are likely to behave over time. He suggests that in the early stages of a country's development the traditional approach towards tendering is frequently used due to requirements from the project's funder, e.g. the World Bank, but that through time the approach may change as the country develops and begins to explore other approaches to procurement, such as design and build, BOT and BOOT type contracts. He suggests that these core complex forms of procurement may incur cost and/or resource implications which the firm may not have predicted at the outset.

Location

Geographic location is obviously an important consideration for a host of reasons, e.g. climate, availability of resources and communications, but Ostler identifies that the proximity of adjacent operations to a new potential territory is a factor worth considering since it may create competitive advantage for the firm. He gives the example of contractors working in Hong Kong obtaining a competitive advantage in the new territory of parts of China due to the proximity to existing operations.

Language

At the outset the question of different languages is clearly one which will be important for the firm. Setting up in a country where the main language is that of the firm's home country is obviously an advantage. However, Ostler suggests that the issue of language may well be cyclical. He points out that very often in the early stages of a developing country's life there is a heavy dependence on the use of multi-lateral aid and international consultants with the concomitant result that English is very often the working language. This, he suggests, brings a comfort factor into operating in a new territory. However, he also suggests that once the country progresses through a few stages of its development a number of indigenous practices emerge and develop with the result that the language priority shifts back to the host country's language.

Market similarity

Market similarity effectively means that markets which are substantially similar to the firm's domestic market in terms of contract terms, contract administration and payment systems may well be more favourable, and less risky, than those which differ substantially. Firms which explore opportunities in countries where the market is dissimilar may face certain cost disadvantages in becoming familiar and therefore competitive in the new market.

Project funding

This factor is one which is at the centre of a firm's considerations when assessing the strategic implications of identifying and pursuing opportunities on an international stage. Issues which must be taken into account when considering this factor include:

- existence of agency funding;
- extent of host funding;
- currency of payments;
- existence of mechanisms for getting the money out of the host country where payments are made, either in whole or in part, locally;

- existence of tied funding restricting the competition to those contractors whose own countries offer funding.

These factors will affect the strategy adopted by the firm.

Legal and fiscal

The firm's overall strategy can be significantly influenced by the prevailing legal and fiscal regimes existing in the host country. Matters such as the basis of law, the history or arbitration in the host country as well as other, perhaps lesser known, dispute resolution methods and the national court procedure must be identified at the outset in order that the extent of legal risk can be ascertained.

The same caution applies to the fiscal regime which may be imposed upon the international contractor. Matters such as: the local tax rate; repatriation of profits; payment of staff and the appropriate income tax issues, must be addressed at the outset.

Client and consultant assessment

It is also important that the firm carry out a detailed assessment on the client(s) and consultants with whom it is likely to be working for and with. This activity cannot be underestimated since the firm's perception of the role of the client and consultants may be very different from what it actually will be.

The overall project structure as well as the structure of the client organisation must be clearly understood. In addition, the roles and responsibilities of the client and the consultants must be clearly documented since the method of working in the host country may well be completely different from that in the firm's own country. A detailed review of the client and the professions involved in the host country's construction industry, their attitudes and methods of working should be obtained via the appropriate trade papers in the host country and via the professional bodies and associations representing such clients and consultants.

The indigenous construction environment

Ostler suggests that this aspect of the firm's strategy should not be overlooked. Many countries' own contractors may well have an international presence themselves and, as a result, may represent competition to the firm seeking to internationalise. On the other hand these indigenous firms may represent potential partners for the firm and may offer the firm in-depth knowledge of the local markets or even become joint venture partners.

Competition

It goes without saying that a detailed review of the firm's competitors likely to be interested in the host country should be undertaken. Each of the factors discussed above would play an important part in the identification and development of a firm's strategy for entry to a particular country.

Ostler has suggested that once all these factors have been identified and analysed for each strategic opportunity the results could be plotted using a weighted index on a Country Attractiveness/Company Strength matrix. The variables discussed above would obviously be taken into account. He suggests that the firm should rank in order of importance those areas which it considers the most important and are representative of competitiveness and country attractiveness. Figure 7.3 is a graphical example of how such a matrix would look.

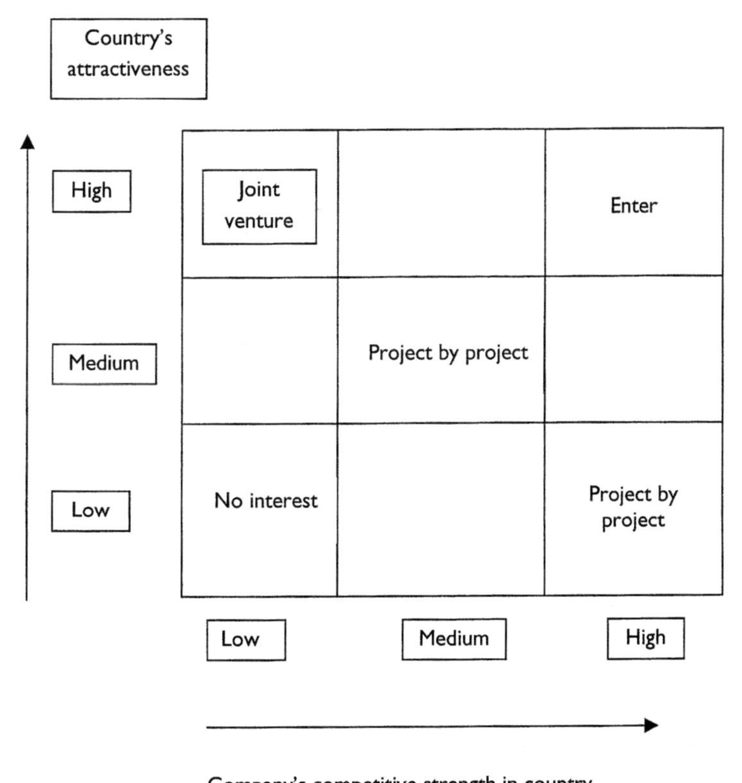

Fig. 7.3 Country attractiveness and country strength.

Ostler has suggested that, while the approach for reviewing strategic opportunities overseas is somewhat subjective, using and analysing the categories previously discussed and by plotting their position on the matrix shown above a relatively good indication of the strategic opportunities available to the firm can be obtained. Two themes underpin the foregoing factors:

- strategic positioning;
- competitive intelligence.

Strategic positioning competitive intelligence

Ngowi & Iwisi [23] suggest that strategic positioning is fundamental if construction firms are to operate successfully in the global market. They also suggest that competitive intelligence is one of the tools that can meet the requirements of identifying and developing a strategic position.

A number of researchers have stated that a business orientation to satisfying the client has been slow to evolve in the construction industry. Porter has identified that marketing is one of the primary activities in the value chain but that in the construction and engineering sector the approach has to be somewhat different from, say, a fast moving consumer goods market.

Following on from the concepts of competitive advantage and differentiation it can be seen that strategic positioning means performing different activities from competitors or performing similar activities in different ways. In the construction industry there are two main generic activities: building and civil engineering. Within these two main groups there are a whole host of generic sub-groups including: housing; retail; offices; industrial; tunnels; railways; power stations, etc. Ngowi & Iwisi [23] suggest that choices of strategic positioning determine not only which activities a firm will perform and how it will configure individual activities but also how activities relate to one another. Essentially, a firm will need to develop a fit amongst its activities and this will, eventually, lead to profitability.

To choose a strategic position a firm should view the global market as the context of operations where highly developed competence in a particular subsector will enable it to pursue opportunities and create some others wherever conditions permit. To create a strategic position, a firm has to allocate adequate resources to the effort and ensure that everyone in the firm is committed to the position created. Once such a strategic position is created the firm has to ensure that such a position is maintained by continuously innovating new features or dimensions in order to stave off potential competition.

Ngowi & Iwisi [23] also point out that a fundamental function of strategic positioning is the firm's ability to identify, collate and analyse global information, i.e. it must possess competitive intelligence. The survival and growth of a firm in its strategic position depends, to a significant extent, upon the quality and accuracy of information and data which it uses to make decision and formulate its strategy.

Essentially, the type of information is known as competitive intelligence. Buckley *et al.* [24] define competitive intelligence as '... a formalised, yet continuously evolving process by which the management team assesses the evolution of its industry and the capabilities and the behaviour of its current and potential competitors to assist in maintaining or developing competitive advantage.' In addition Gilad *et al.* [25] define competitive intelligence as '... the activity of monitoring the environment external to the firm for information that is relevant for the decision-making process in the firm.'

The use of competitive intelligence may be crucial in assisting the firm identifying and analysing those factors which underpin and affect the issues, discussed earlier which should be taken into account when developing a strategy for internationalisation.

Gilad [25] has identified five main elements which are crucial to the development and execution of effective competitive intelligence. These include:

- collection – data are gathered relevant to the matter under consideration;
- evaluation – the data are ranked and their accuracy, timeliness and usefulness are determined;
- storage – the data are organised and maintained;
- analysis – where the data are reviewed, tested and subjected to challenge;
- dissemination – where the data are used for decision-making.

In addition to these elements, Gilad [25] has identified that such data originate from two main sources: published sources and field sources. Both these categories are further defined as: what others say about competitors and what competitors say about themselves. Under the published category information relating to 'what others say about competitors' may be found in the following:

- books;
- periodicals;
- trade press;
- research publications;
- technical journals.

Under the field sources category information relating to 'what others say about competitors' may be sourced from the following:

- customers;
- suppliers;
- subcontractors;
- agents;
- clients.

Information on 'what the competitors say about themselves' can be drawn from their own corporate literature, e.g. annual accounts, corporate newsletters, press releases, seminars, etc.

The importance of the role played by competitive intelligence cannot be underestimated: the factors which must be considered when assessing strategic opportunities must be subjected to the rigours of competitive intelligence. Leaving such matters to chance means that any strategic decision by the firm is more likely to be subjective rather than objective.

This chapter has investigated a number of issues relating to international construction. The discussion on the background to the practice of international construction set the scene for engendering a basic understanding of why, in the past, international construction has been so important as well as giving an indication of what forms it took. Moving on from this historical perspective into the present, the chapter has examined the contemporary profile of the size and structure of the international construction market by considering the size and geographical dispersion of the international market, the share each particular country has of the market, and the share each sector of the construction market has, i.e. civil engineering, building, process engineering, etc.

By using this foundation knowledge the chapter goes on to analyse the main reasons why a firm would wish to internationalise. In addition to providing a brief analysis of some of the historical reasons the chapter examines a wide variety of reasons why firms seek to internationalise. Some of these reasons are micro-orientated, i.e. they originate within the firm itself, e.g. to maintain shareholders' return, to balance growth and to increase turnover. A number of other reasons are macro-orientated, i.e. they are in response to some outside influence, e.g. favourable tax advantages, political stability of the host country, the existing and potential economic growth of the host country and the level of indigenous competition.

No discussion of international construction would be complete without an analysis of the characteristics and obstacles to internationalisation. The chapter identifies some of the main characteristics of internationalisation, e.g. the existence of a fragmented industry structure, geographical dispersion, payment issues and profit repatriation. These characteristics are then used to provide a detailed review of the obstacles that make internationalisation problematic.

The chapter then moves from a theoretical position to a more applied emphasis by going on to discuss competitive advantage and strategy in international construction. A great deal of discussion and analysis of Porter's model and how it applies in the international construction setting takes place, i.e. a discussion of the factor conditions, domestic demand conditions, related industries and the firm's overall structure and competitiveness. Building upon this analytical approach the chapter ends with a detailed review of the factors, issues and criteria which must be considered when identifying and developing an internationalisation strategy.

References

[1] Linder, M. (1994) *Projecting Capitalism. A History of the Internationalisation of the Construction Industry*. Greenwood Press, Westpoint, CT.

[2] Cannon, J. & Hillebrandt, P. (1991) Diversification. In: *The Management of Construction firms: Aspect of Theory*. J. Cannon & P. Hillebrandt (eds), pp. 31–43. Macmillan, Basingstoke.

[3] DETR (2000) *Construction Statistics Annual*. The Stationery Office, London.

[4] *Engineering News Record* (2000), 14 August. McGraw-Hill, New York.

[5] Bauml, S. (1997) Engineering and construction: building a stronger global industry. *J. of Management in Engineering*, **13**, No. 5: 21–24.

[6] Strassman, P. & Wells, J. (1988) *The Global Construction Industry: Strategies for Entering Growth and Survival*. Unwin Hyman, London.
[7] Seymour, H. (1987) *The Multi-national Construction Industry*. Croom-Helm, London.
[8] Maloney, W. & Hancher, D. (1997) *Strategic Positioning for ASCE*. Construction Congress V, Minneapolis, MN.
[9] Langford, D. & Rowland, V. (1995) *Managing Overseas Construction Contracting*. Thomas Telford, London.
[10] Limerick, Earl of (1980) Selling construction. In: *International Construction*, S. Urry & A. Sherratt (eds), pp. 1–8. Construction Process, Lancaster.
[11] Imbert, I. (1990) Human issues affecting construction in developing countries. *J. of Construction Management and Economics*, **8**, No. 2: 21–32.
[12] Crossthwaite, D. (1998) The internationalisation of British construction companies 1990–96. An empirical analysis. *J. of Construction Management and Economics*. **16**, No. 4: 389–395.
[13] Jauch, L. & Glueck, W. (1985) *Business Policy and Strategic Management*. 4th edn. McGraw-Hill, Singapore.
[14] Male, S. (1991) Competitive advantage in the international construction industry. In: *Competitive Advantage in Construction*, S. Male & R. Stocks (eds). Butterworth Heinemann, Oxford.
[15] Wang, S., Tiong, R., Ting, S., Ashley, D. (1999) Political risks: analysis of key contract clauses in China's BOT project. *J. of Construction Engineering and Management*. **125**, No. 3.
[16] OECD (1992) Globalisation of Industrial Activities. Four Case Studies: Autoparts, Chemicals, Construction and Semi-conductors. OECD, Paris.
[17] Langford, D. & Zawdie, G. (2000) The influence of culture on the internationalisation of construction. Proceedings of CIB TG29. Construction in Developing Countries, Botswana. December.
[18] Flanagan, R. (1994) The features of successful construction companies in the international construction market. In: *Strategic Planning in Construction Proceedings of the AT*. Ethan International Seminar on Strategic Planning in Construction Companies. Haifa, Israel.
[19] Male, S.P. & Mitrovic, D. (1999) Trends in World Markets and LSE Industry. *J. of Engineering Construction and Architectural Management*, **6**, No. 1, 7–20.
[20] Porter, M. (1985) *The Competitive Advantage of Nations*. Free Press, New York.
[21] Porter, M.E. (1990) *The Competitive Advantage of Nations*, p. 127, Fig. 3.5. Macmillan, London.
[22] Ostler, C. (1998) *Country Analysis: Its role in International Construction Industry Strategic Planning Procedure*. Conference on International Construction Marketing, Leeds.
[23] Ngowi, A. & Iwisi, D. (1998) *Strategic Positioning in the Construction Industry*. Proceedings of the International Marketing Conference: Research into Practice, Leeds.
[24] Buckley, P., Pass, C., Prescott, K. (1992) *Servicing International Markets – Competitive Strategy of Firms*. Blackwell, Oxford.
[25] Gilad, B., Gordon, G., Sudit, E. (1993) Identifying groups and blind spots in competitive intelligence. *Journal of Long Range Planning*, **26**, No. 6: 107–117.

C Techniques for the strategic planner

8 Portfolio management, Delphi techniques and scenarios

Business portfolio management

Portfolio management comprises a set of techniques which are often used by strategic planners to integrate and manage strategically a number of subsidiaries, often operating in different industries, that comprise the corporate whole. The larger the business the more likely it is there will be a number of operating units in existence which need to be integrated and managed strategically. One of the methods for achieving this which is most often discussed in strategic management literature is product market portfolio analysis [1], [2]. Its use is primarily discussed in terms of large, diversified companies that have to consider many different businesses or strategic business units (SBUs), with different products on sale in the market place or under development. In such a situation the strategic planning process can become complex and the main concern is to ensure a balanced range of businesses or activities [3]. In order to provide a structure and subsequent guidance for decision-making under these conditions a number of different techniques have been developed, using the same form of matrix analysis. Underlying this analysis using portfolio techniques are the concepts of the experience curve and the product life cycle discussed in Chapter 4. McNamee [1] indicates that portfolio management necessitates the determination of three fundamental characteristics of a product's or SBU's strategic position:

- Its market's growth rate;
- Its relative market share in comparison to the market leader;
- The revenues generated from the product's sales or the SBU's activities.

The Boston Consulting Group growth–share matrix

The fundamental characteristics of portfolio management can be represented pictorially. The Boston Consulting Group (BCG) growth–share matrix (Figure 8.1) is a widely reported example of this. The four cells of the matrix represent different decision situations.

Cash cows

These are the products or SBUs that generate more cash than they require to operate. They will have a high market share but in a low-growth market. As

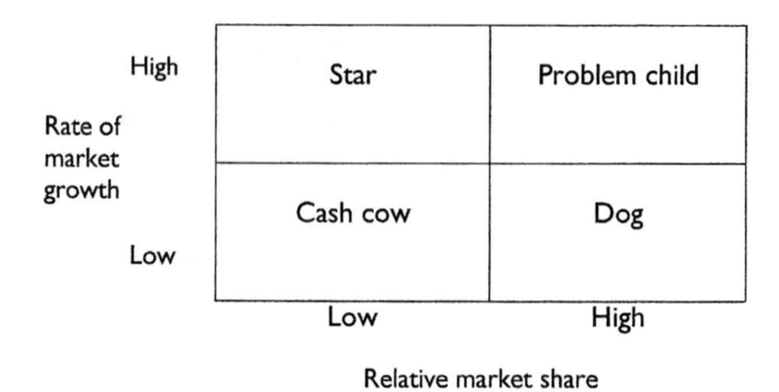

Fig. 8.1 BCG growth–share matrix.
Source: Tools and Techniques for Strategic Management, McNamee [1]

they are cash generators this money can be utilised in other parts of the business.

Problem children or question marks

These activities require net cash inputs and are usually in a growing market where market share should be maintained or increased. The dilemma for the strategist is that the market is growing rapidly, the product or SBU has a low market share but requires substantial cash injections to allow market share to be built and thus allow the activity to become important for future profitability. The development of a project management service is an example of such an activity in the construction industry.

Stars

These activities have high market shares in fast growing markets and generate considerable amounts of cash. However, in order to maintain this position they also require considerable amounts of investment. Cashflow is, therefore, in approximate balance. Over time these activities could move into other quadrants on the matrix.

Dogs

Dogs have low market share in a low-growth market. McNamee [1] and Wheelan & Hunger [4] refer to two types of Dog, the former using the notion of 'kennel' to describe this situation. The 'genuine Dog' normally requires divestment. However, the 'cash Dog' may need careful management to generate some cash.

The BCG matrix has been criticised on a number of points which are outlined below (for further details see Jauch and Glueck [5], McNamee [1], Wheelan & Hunger [4].

- The two-dimensional nature of the matrix is simplistic and the use of relative market share and growth rate may reflect only partly the determinants of a product's or SBU's success.
- The postulated strong link between market share and profitability may not always hold.
- High growth markets may not always be the best to enter. For example, project management has been suggested as being a high-growth market in construction [6]. However, there are considerable risks attached, not least that of liability. Substantial investment is required if an organisation is serious about offering the service and, in addition, personnel require a different orientation and set of management skills from many professions that purport to offer such a service. Potential entrants may therefore face high entry barriers.
- The main reference point for the analysis is the market leader and no attention is paid to other fast growing but small enterprises.
- The matrix analysis does not represent adequately those businesses that are growing rapidly or are in a new industry in the business start up or growth stage.

In view of the limitations of the BCG growth-share matrix a number of other matrix-type analytical techniques have been developed, for example, General Electric's 9-cell industry attractiveness business screen; the 12-cell product/market evolution matrix; the 4-cell BCG strategic environment matrix [1], Wheelan & Hunger [4]. Johnson and Scholes [3] argue that portfolio analysis, using the preceding techniques, should not be viewed as a comprehensive method for evaluating different strategies but should be placed in the context of a preliminary step in raising the general awareness of managers of strategic issues. The following section relates the relevance of portfolio analysis to the construction industry and introduces the idea of 'project portfolio management'.

Portfolio management and the construction industry

Two important issues have to be emphasised with respect to the application of the above concepts to the construction industry. First, many of these tools or techniques have been developed from a manufacturing industry base where the methods of pricing and design/production processes differ considerably from those of contracting. Second, in large, diversified construction related firms that have subsidiaries operating in a traditional manufacturing type of environment or those conglomerates with construction or property development subsidiaries, the use of portfolio analysis tools has an application at group or headquarters level [7]. However, the applicability of these ideas in the construction industry is compounded by the fact that the product in construction can be viewed as either (i) a completed project in the form of a building or other type of facility (in many cases a one-off for a particular client) or (ii) as a service. If the latter is true firms in construction may offer a portfolio of services. In the case of a contractor a portfolio of services may comprise design and build, management contracting and general contracting. In the case of a quantity surveying consultancy the portfolio of services could include feasibility studies, cost planning, production of tender documentation and post-contract final accounts and finally, project management. Additionally, there are problems surrounding the definition of markets in construction and subsequently in defining market share [8].

For many organisations operating in the construction industry, such as a medium-sized regional contractor or a consultancy firm, the issue may not be one of managing different subsidiaries but the strategic management of different types of project. However, the management of different types of project also has implications for the larger organisation in construction.

The management of a portfolio of contracts can be crucial to the success of the contracting company. Ball [9] argues that profitability is dependent on achieving a balanced mix between projects that are under way and those being bid for. He sees the ability to diversify along the project dimension as a scale economy in the industry. Other advantages of project portfolio management include:

- The ability to enter markets easily without fear of retaliation. An issue of considerable importance in the manufacturing industries [10].
- Access to new clients, tender lists and/or contacts through acquisition of other firms' project portfolios.
- Spreading risks over a number of projects to reduce the impact on turnover, profit and/or company operations of any one project. However, the nature of construction projects and the inherent risks attached may mean in practice that it only takes one project to destabilise a company's operations. For example, Morrell [11] discusses in detail the impact of the problems encountered on the Kariba Dam that contributed to the eventual liquidation of Mitchell Construction.
- Market withdrawal is easier, especially for larger firms, since the commitment to any single market may be for the duration of one project only.
- Increased bargaining power with clients can be obtained due to a wider client, and hence profit, base.

The applicability of project portfolio management is not restricted to the contractor but also extends to architectural, surveying or engineering consultancies in the construction industry, especially now that competitive fees bidding is common. Unlike the contractor these organisations are not involved directly in the physical construction of a facility through the on-site production process.

However, each of these types of firm derives its method of working, organisational structure and long-term business direction partly from the fact that a diversity of clients procure services on the basis of individual client project requirements. Therefore, many of the points highlighted above apply equally to consultancy organisations in construction.

The application of portfolio analysis to the selection of projects can be extended to apply to the range of businesses that a company seeks to complete. Using the current portfolio as a study point the company has a range of options to change the portfolio through organic growth (e.g. Morrison Construction which grew from a £64m company in 1990 to have a turnover of £300m in 2000 before being sold to Anglia Water).

Alternatively companies can form joint ventures to expand turnover or acquire new companies or merge with others. Examples of this strategy have been observed in respect of an Anglo, Japanese, and Chinese joint venture in the building of Hong Kong's Chap Lap Kok airport and HBG's acquisition of several regional companies in the UK including Kyle Stewart, GA Construction and Higgs & Hill.

The portfolio of activity can expand by doing one or several of the following.

Internally driven expansion

- Expand into similar markets to these already served (e.g. a move from building speculative green field housing to refurbishing 'lofts' for a younger market).
- Market development (e.g. investments to ensure a larger share of an existing market – e.g. investment in capabilities to compete in the PFI health market).
- Product development (spreading an idea over a wider range of products – an architect spreading out design capabilities to say furniture as well as construction design).

Externally driven expansion

- Horizontal integration by expanding into related markets by merger or acquisition (e.g. the putative but eventually abortive merger between Bovis and W.S. Atkins).
- Diversification by expansion into different markets (e.g. Aukett Associates acquired an architectural practice in Prague thus accessing a European network.

The quest for expansion needs to be based upon some forecasts of which areas are set to expand and which others are in decline. Companies involved in strategic management structure their exploration of the future by using known techniques for probing for underlying trends. The techniques used here are Delphi surveys and scenario planning.

Delphi techniques

In Greek mythology the god Apollo is afforded a special place for he was said to be able to foretell the future. The followers of Apollo built a temple dedicated to his worships at Delphi, some 170 km north-west of Athens. What was special about Delphi was that fumes escaping from volcanic fissures were inhaled by acolytes who fell into trances and then made utterances interpreted by priests as predictions for the future. What has this to do with corporate planning? The drinks cabinet in the board room may be well-stocked and several directors may go into trances on occasion, but history can be of use. During the 1950s the RAND Corporation developed a technique for eliciting expert opinion by questionnaire and this survey technique was named the Delphi technique. An early user of the Delphi technique was the US military who asked experts to forecast where nuclear attacks would take place. By 1964 more peaceful uses had been found, for a 'Delphi' was used by the US Government to try to predict the future of science and technology. Now Delphis are used in business to predict the future of an organisation.

The Delphi technique can best be explained by using an example. Let us take a construction company which is trying to assess what the construction environment will be like in 2020. The first step is to assemble a number of experts in the company – not only directors but functional specialists such as accountants and engineers. When these participants have been briefed, the following question can be asked of them.

Question 1: In the current study a period of thirty years is being considered. It is possible that inventions not yet visualised could have a significant impact on the construction industry over this period. It has been seen that the industry has changed rapidly over the last ten years both in the technology and contract systems. What changes can you foresee in the next thirty years? List below the main breakthroughs that you think are urgently needed and which are feasible within the next thirty years.

A selection of items which may be submitted in response to this question are:

- The nature of the environment
  Sub-aquatic cities
  Building in space
  Building for space settlements
  Subterranean cities
- Ownership and contractual matters
  Design, build, maintain and refurbish contracts
  Industrial conglomerates having their own direct labour organisations (i.e. vertical integration including construction services)
  Contractors required to put equity into projects

Builders' merchants become major equity stakeholders
Totally integrated construction services
No divorce of design and construction

- Construction techniques and technology
  Throwaway buildings (5-year life span)
  Entirely robotised construction
  Lightweight 'clip-on' materials
  Changeable environment (i.e. visualise a different environment and it will be given. For example, I need a break from work and I can organise my environment so that 'visually' I am somewhere else).

The next phase is to test the consensus of opinion amongst the expert panel. The suggestions can be put to them in the form of a question. Thus:

Question 2: Listed below are the breakthroughs suggested by the panel. Indicate as a percentage mark the probability of each breakthrough being implemented by the period of time indicated.

A completed questionnaire is shown in Table 8.1.

Following the panel's assessment of the projects the data collected can be analysed, based on a rolling forecast of when the events or possibilities are likely to occur. This can be presented in diagrammatic form as shown in Figure 8.2.

Table 8.1 A completed Delphi questionnaire

	% probability of event occurring in year					
	2001	2005	2010	2015	2020	2025
1. Environment						
Sub-aquatic cities	0	5	15	15	20	30
Building in space	5	10	40	80	100	100
Building for space settlements	25	100				
Subterranean cities	50	100				
2. Ownership and contracts						
Design, build, maintain & refurbish contracts	100					
Industrial conglomerates having their own DLOs	20	10	5			
Contractors' equity	0					
Builders' merchants have equity stake	10	10	10	10	10	10
Totally integrated construction	40	50	60	70	80	90
3. Construction techniques						
Throwaway buildings	2	80	90	90	90	90
Robotised construction	10	90	100	–	–	–
Lightweight materials	50	80	100	–	–	–
Changeable environment	0	0	0	0	0	0

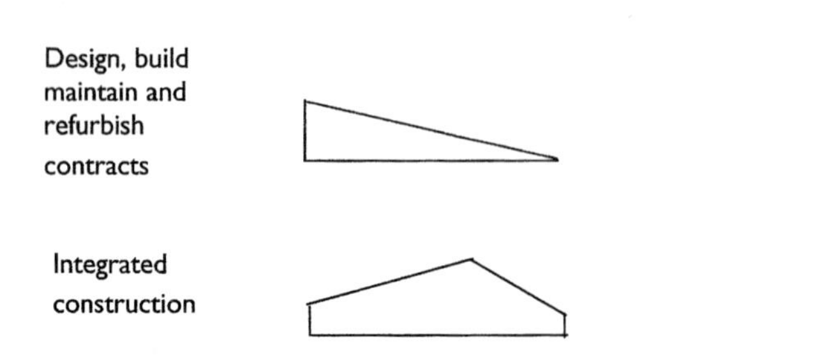

Fig. 8.2 Probability of events occurring.

The probabilities supplied by the respondents are averaged and the highest probability of a date is signalled with the shortest time horizon being located earliest in the diagram. As can be seen the presentation appears as a bar chart with the events considered to be the earliest possibility appearing first. This gives more of a sense of timing enabling adjustments to the possibilities of future events. The taller the apex on the triangle the more confident the reports are that this is likely to happen and the actual apex indicates when it is predicted to occur.

This approach is most useful when a problem does not lend itself to analytical techniques and is consequently subject to personal judgement. Also the approach is freed from the use of historical data as a guide to the future. By identifying and separating out the experts for this kind of research one can avoid the phenomena that often beset meetings of such people – meetings dominated by bandwagons, formal status divisions or strong personalities. The possible future environment for construction can thus be predicted, the process can be used as a communication tool and the results as a mechanism to tease out the pros and cons of various policy options.

In the example given a general approach has been used with 'experts' giving an opinion on all matters but obviously specialist groupings can be formed. Frequently these groupings are based around operating divisions within the company. For example, after an environmental framework has been established by a general 'Delphi' then specialist groups could consider futures in, say, house-building, international projects, industrial or commercial sector building, etc.

The Delphi technique is not without its difficulties. In many trials it has been found that the longer the planning horizon the greater the variation in the predicted date of an event. This position will be familiar to construction project planners who at the outset of a project are able to project the time of completion of foundations with greater certainty than the topping out ceremony. The time between the foundations and the topping out can encompass many events, which make the date of the topping out uncertain. Also, in using this technique, it has been found that there is a tendency to be pessimistic over the long term but optimistic in the short term. This tendency is associated with a sense of amortising the future. Today's problems are those needing attention and tomorrow's problems do not assume much importance. Also respondents show a preference for simple and certain futures. In themselves these factors need not matter, for the aggregated opinion of the experts can provide useful insights for corporate planners and form a useful part of their techniques.

A recent example of a Delphi undertaken at the level of whole industries was the Technology Foresight programme undertaken in the UK in 1995. This programme was intended to focus upon the technological futures for selected industries and the construction industry was the subject of a specialised study. The starting point of the Delphi survey was to identify technological and other factors which could improve future competitiveness of the construction industry. The study was conducted in the classical manner; stage 1 collected together a group of experts who were questioned about the future for the industry. The data from those discussions were then aggregated and presented to the group of experts who were invited to consider their views in the light of information from other experts.

From the analysis a set of 80 questions emerged which were grouped under the general headings of:

- Buildings in use
- Changing client demand
- Design
- Finance and funding
- Land use
- Materials
- Productivity improvements
- Quality of life
- Regulations, safety and security
- Sustainable development.

Each of the above theories was evaluated against several issues, which included wealth creation, the impact on quality of life, the comparative position of the UK in respect of European and international standards and the constraints on the likelihood of events happening. The data were collected from 196 experts. The items which were seen to have the most positive influence are shown in Table 8.2.

The outcomes for the Delphi study presented a future of the industry which was bound by the conventional economic forces and the growing international consensus that modern societies are liberal and capitalist in political and economic orientation. Perhaps this is not surprising since the respondents to the questionnaire will have likely been corporate men (including the two authors

Table 8.2 Leading factors seen to have impacts on the future wealth creation and quality of life

Area of impact	Leading factors having an impact
Wealth creation (positive effect)	The balance of trade in construction materials, components and services. More out of deficit and into credit (i.e. we should export more construction or reduce imports of construction to improve wealth creation
Wealth creation (negative effect)	Increases in planning legislation and environmental taxes
Quality of life (positive effect)	Travel times to work halved
Quality of life (negative effect)	Buildings are designed for short life spans and with a limited function

of this book!); the voices of women, young people and those who were not 'on-side' with the establishment are unlikely to have been heard. Indeed, at one of the regional workshops of the 30 people in attendance there was one woman, very few people under 40 and one of the authors was the only person wearing apparel other than a dark, business suit! However it must be pointed out that the workshop was held in Dundee!

Scenarios

The following exchange was said to have taken place at a Cricketers' Association (the professionals 'trade union') meeting. In a contribution to discussion Phil Edmonds of Middlesex and England is reported to have said, 'Let's postulate a hypothetical scenario'. The audience looked quizzically at him and David Lloyd then of Lancashire, noted *sotto voce*, 'It's all right, lads, he just means "let's pretend"'.

Is scenario planning just a fancy name for 'pretending'? If so, it has little value to the corporate planner, but scenario planning is more than pretending. In the words of Wilson [12], it is an 'exploration of an alternative future', for, if we make certain assumptions, then the future may take a certain direction. Scenario planning can be described as follows:

(1) Hypothetical – it exposes us to the numerous possibilities of the future.
(2) Vague – it does not give detailed descriptions but provides a generalisation of what may be.
(3) Multi-disciplined – it is an overview which brings together all aspects of a society be they technological, economic, political, social, demographic or environmental. The scope of a scenario will depend upon what is attempted – a global scenario will be wider in the impacts it considers than a national scenario. Equally an industry scenario will be wider than one for an individual firm.

This section will focus upon the particular use of scenarios in companies and construction organisations. An essential building block of a scenario for companies and construction organisations is the 'trend', which was discussed in Chapter 4. Trends are naturally isolated from one another, each trend giving one set of information, but the scenario binds these into a complete, if hazy, picture. A scenario can be built and used as a framework for corporate planning, on which a construction organisation will base its ability to adapt to a future changed environment. A scenario attempts to portray the future shape of the construction world and how the company can adapt to fit it. In a sense this is corporate ecology – the firm evolves to suit the environment in which it lives.

In order for the firm to 'fit' with the business environment it must have some understanding of the current and future business environments. The technology foresight [13] construction industry report is again helpful in aiming towards an understanding of political, economic, social and technological factors which will shape the business environment. The foresight panel presented a utopian scenario which could be summarised as being characterised by a growing economy which is based upon principles of sustainability and populated by extroverted confident and participative people who are individually rewarded. The foresight report also recognises the possibility of a dystrophic future where economic growth is stinted, nationalism abounds and peoples and governments are slim of confidence, defensive and introverted.

As ever we are not likely to be blessed with all of the upside but neither be cursed with all of the downside – perhaps just enough to make us feel that all of the worst is happening. This responsiveness to new environments implies a level of passivity on behalf of the firm yet planning suggests some level of proactivity. Not only does the construction organisation react to the world but it can also assist in shaping that world.

Whatever the purpose of the action which follows a scenario it is essential to have a planning framework from which to work. The objectives of scenario planning for companies are therefore:

- To set up forecasts of alternative economic futures;
- To set up alternative visions of the future;
- To identify branching points in the future which can act as early warnings for the company.

These objectives set the context for testing how well a company's products or services fit the various postulated futures. More specifically planners can test the outcomes of various company strategies in different future environments. How well the company can exploit the future environments is a function of how well the individual firm can fashion distinctive competencies and so create a competitive advantage. It may be argued that a distinctive competence is harnessed through an understanding of how a society or an individual market is changing. If the firm can 'surf' the wave of change that it has spotted then it can be better placed to compete or win work in other ways. Van der Heijden [14] expresses the connection between the business environment, distinctive competence and competitive advantage as the generic business idea. This is shown in Figure 8.3.

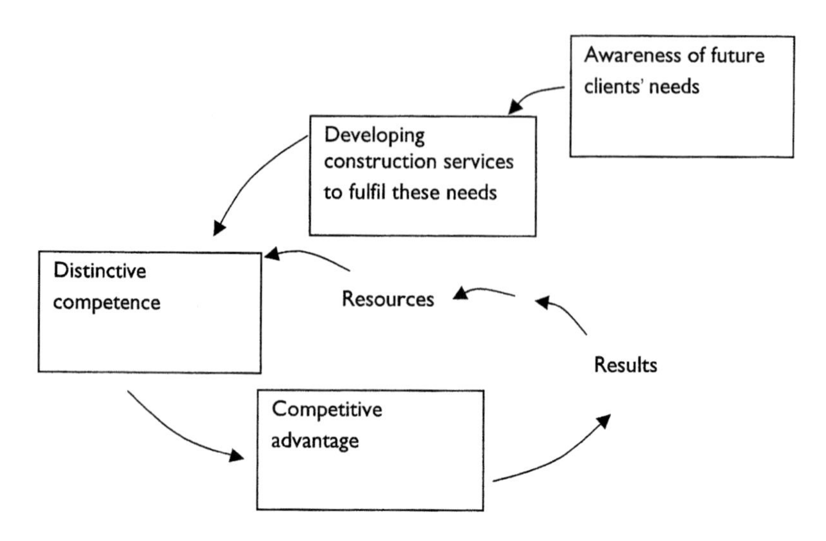

Fig. 8.3 The generic business idea (after Van der Heijden).

Van der Heijden goes on to customise this generic model. In one of his examples a construction company is depicted. Interestingly he sees the world of a construction company through different eyes from most of those with a background in construction. The interpretation of the construction company is centred upon a client's perspective. Clients want a product (a building) to give them service for a long time, they buy before the product can be inspected and so are risk averse. In this business environment the reputation of delivering a quality product is of central importance. This quality is underpinned by formal quality assured buildings delivered to satisfied clients. The firm demonstrates to clients its reputation by solipsically referring clients to what Van der Heijden calls the 'installed-base'. This portfolio or 'installed-base' acts as a barrier to entry for new players. Competition is also stifled by presenting the firm as a collaborative business partner as opposed to a spiteful adversarial predator. If the company is to succeed in generating work by partnering or negotiating rather than tendering it must have a culture based on collaborative relationships and investment in training and development of the people within the firm and flexible business arrangements between firms. These factors play out in the model shown in Figure 8.4.

In order to benefit from such models the firm needs alternative views of the future. The model used in Figure 8.4 models 'quality' as a defining idea by which reputation is built and early boarders of the quality bandwagon will have benefited. However, the central truth is that the power of a competence (in

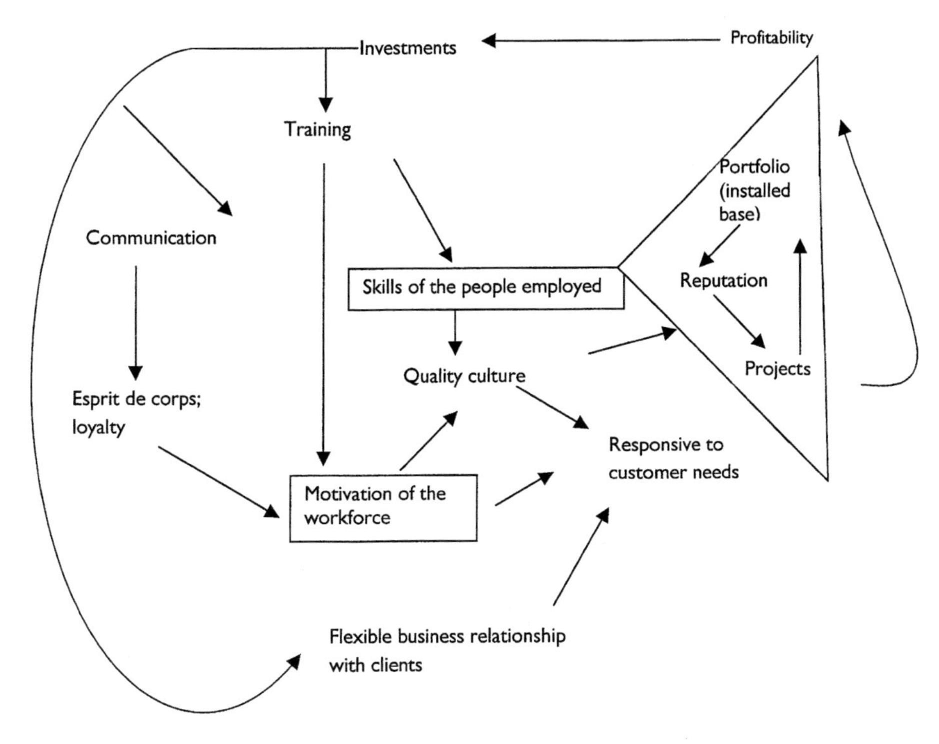

Fig. 8.4 The business idea of a construction firm (after Van der Heijden).

the example in Figure 8.4 – quality) is likely to wither as others seek to replicate the idea. The nimble company seeks advantage by being able to detect the next driver of change and this means that the firm must have staff whose function is to regularly conduct a *tour de horizon* to create fresh scenarios and thus detect trends which may shape the future. These trends can be modelled on two axes: the probability of a trend occurring and the malleability of a trend. Figure 8.5 demonstrates the mapping of trends on these axes using four examples.

(1) *Population and demographic trends*
It may be that planners predict urban regeneration in the North of England with population shifts to this area. The planners see this as a high probability but can do little to shape the direction of this trend.

(2) *Law and legislation*
Planners may envisage changes in, for example, employment law and assess that the probability of these occurring is high. The firm can insulate itself from the effects of the changes by making certain strategic decisions, for example, sub-contracting all work.

(3) *Technical developments*
Imagine that planners foresee that by 2015 the motorway and the conventional car will have been replaced by electro-mechanical cars which clip

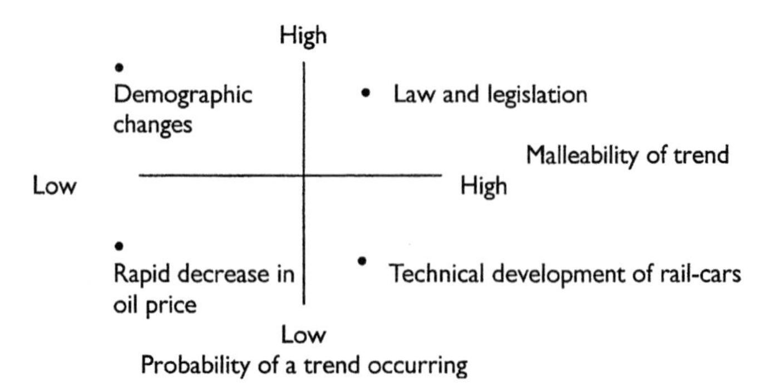

Fig. 8.5 Some trends mapped.

into an electrical rail fitted on a motorway lane. The rail governs speed and spacing between the cars. If drivers wish to leave the motorway they slip off the electrical rail and return to 'petrol power'. There may be a low probability of this happening but a firm could respond to this technical change and shape it through its own research into electrical transport systems.

(4) *Increase in oil prices*
This may be a possibility and it would certainly have an effect on a company's workload, however, it is an event which is outside the influence of the individual construction organisation.

By mapping the trends which can influence the construction organisation the scenario process identifies the initiatives that are needed and seeks to minimise the responses that corporations have to make. This thinking is based upon the formal planning logic that proactivity is preferable to reaction.

The process of formulating scenarios

The formulation of scenarios can be explained in more detail by describing a typical process, taking an international construction organisation as an example. The first task is to select the areas about which scenarios need to be developed. The firm in our example is interested in the following global issues:

- The geopolitical balance of the world
- World economic development.

In the domestic market, however, it prefers to consider:

- The social framework of the future
- Political direction
- Economic futures

- Technological prospects
- Labour and demographic trends
- The legal framework for business.

The scope of the scenario is now defined. The second step is to historically review each issue with particular interest focused on recent history. This isolates the current trends in each issue. In our example the firm may see the world balance of economic order changing to redress imbalances between the wealth of the northern hemisphere and the poverty of the south. Consideration of the geopolitical issue may project *glasnost* to suggest a significant political integration of the world. This review leads to the third part of the process in which the future is projected from the detectable trends with 'probable' changes in each sector being postulated.

Thus far a picture is emerging but the planner has yet to assess the key change points which break away from the 'probable'. For example, if our scenario is for greater economic equality in the world, what are the key break points or factors which will decrease this trend? Finally, the planner has to assess the impact of these trends on the company. If we continue our example, the planner may suggest a strategy to meet expected construction demands in developing countries.

The steps taken to formulate a scenario generate a massive amount of information and some weeding out of the information is necessary to enable it to be manageable. A simple way of filtering the information is to quantify the probability of an event occurring and multiply it by the importance of the event to the company. Thus:

Significance of a trend = probability of an event × its importance to the company

The weighting of trends is crucial; an event may not be considered as highly probable but if it does happen then its effect may be catastrophic for the company. For example, if a mainstay of a company business portfolio is quarrying and processing materials for tarmacadam and a technological breakthrough, such as personal hovercrafts, enables personal transportation to operate without roads, then this would be catastrophic, even though the likelihood of the event is small. However, some events may be set aside altogether as not being influential over the firm.

The filtering process should enable the top fifty trends to be isolated. From here planners move on to use the technique of cross impact analysis.

Cross impact analysis

Events in history are rarely based around single issues although it may be convenient to presume so and say that A caused B. However, life is seldom so simple. Consider, for example, the events surrounding the start of the First World War. Schoolchildren are probably taught that the war started because an obscure archduke, Ferdinand, was assassinated in a remote Balkan state. This single event was said to have triggered the war, but more complex, underlying issues could also be said to have been at work in the struggle for international markets among European nations. More recently, the tower blocks which were built in the 1960s were able to be built because of a fusion of technical capability, political need and encouragement, social needs and acceptance of such blocks, etc. In short, one event is an interaction of many others.

This technique can be illustrated by the example of a construction firm with a strong interest in the timber-framed housing market. Over the next twenty-five year period the corporate planners of this company isolate the following trends:

(1) Acid rain has an increasing effect upon the tree crops in Canada and Sweden.
(2) Canada and Sweden look to restricting the export of timber.
(3) Legislation introduces statutory requirements to improve fire and sound insulation in timber-framed houses.
(4) The decline in the timber-framed housing market is to continue.

Obviously there may be some links between these possible events. For example, acid rain makes restrictions on exports more likely; the decline in the popularity of timber-framed housing makes legislation less likely for there may be no purpose in legislating for a small market. These events may be classified so that several different kinds of relationships can be seen.

- Unrelated – one event occurring does not influence the probability of the second (for example, acid rain and legislation).
- Enhancing – the occurrence of one event increases the probability of a second event either by enabling it to occur or by stimulating its occurrence (for example, acid rain and an export embargo).
- Inhibiting – one event inhibits another (for example, a fall in timber-frame sales and legislation).

However, it must be pointed out that the relationships established go beyond a simple pairing and that repercussive events may be set up. Also the strength of the relationship is important. In our example the relationship between acid rain and timber export embargo seems more likely than that between timber-frame house sales and legislation.

The impact of these different events can be analysed by setting up a matrix to demonstrate the way the probabilities work together. Such a matrix is illustrated in Table 8.3.

What are the chances of events 1 and 2 occurring together? We assign a probability of one event happening, for example, there is a 60% chance of acid rain affecting the timber crop by 2000. If acid rain is going to reduce the crop

Table 8.3 Cross impact analysis matrix

Probability of events occurring by 2010	1	2	3	4
1. Acid rain	0.60	–	0.24	0.30
2. Timber export embargo	0.30	0.18	0.12	0.15
3. Legislation re improved safety standards	0.40	0.24	–	0.20
4. Reduction in timber-framed housing market	0.50	0.30	0.20	–

then an embargo on timber exports seems more likely. The probability of both happening can be calculated by multiplying together the individual probabilities as shown in the table.

To follow this through, reduced timber exports may increase the price of timber which in turn may be reflected in the cost of timber-framed housing. This will influence the price and depress the market. So the export embargo and fall in sales of timber-framed houses are connected by more remote events. The probabilities of all the events occurring can also be calculated. The probability of an export embargo being caused by acid rain is 0.18. The matrix can be then redrawn to show the relationship between the other variables. This is cross impact analysis.

After looking for any cross impacts, planners can proceed to write up the different scenarios. According to Zentner [15] the scenarios presented must be:

- Credible (if not, planners have difficulty in developing strategies)
- Useful (relevant facts)
- Understandable (presented in a clear way).

In constructing a scenario for the construction organisation it seems likely that the main elements to trend analysis would be:

- Demography – where people will live and work.
- Lifestyles – how people will live and work.
- Transport – how people will move around, where to and where from.
- Environment – what kind of built environment will people have?
- Technology – what technological advances can society be expected to make?

From such scenarios one can undertake sophisticated corporate planning in which 'what ifs' can be addressed. This imprecise science is important in setting up different potentialities for the company so that it can structure its operations to match likely scenarios of the future.

References

[1] McNamee, P.B. (1985) *Tools and Technique for Strategic Management*. Pergamon Press, Oxford.

[2] Howe, W.S. (1986) *Corporate Strategy*. Macmillan, London.

[3] Johnson, G. & Scholes, K. (1988) *Exploring Corporate Strategy*, 2nd edn. Prentice Hall, Hemel Hempstead.

[4] Wheelan, J.D. & Hunger, T.L. (1987) *Strategic Management*, 2nd edn. Addison-Wesley, Reading, MA.

[5] Jauch, L.R. & Glueck, W.F. (1988) *Business Policy and Strategic Management*, 5th edn. McGraw-Hill, Singapore.

[6] MAC (1985) *Competition and the Chartered Surveyor: Changing Client Demand for the Services of the Chartered Surveyor*. Report by Management Analysis Centre for the Royal Institution of Chartered Surveyors, London.

[7] Ramsey, W. (1989) Business objectives. In: *The Management of Construction Firms: Aspects of Theory*, P.M. Hillebrandt & J. Cannon (eds), pp. 9–29. Macmillan, Basingstoke.

[8] Hillebrandt, P.J. (1984) *Analysis of the British Construction Industry*. Macmillan, London.

[9] Ball, M. (1988) *Rebuilding Construction. Economic Change in the British Construction Industry*. Routledge, London.

[10] Porter, M.E. (1980) *Competitive Strategy. Techniques for Analysing Industries and Competitors*. Free Press, New York.

[11] Morrell, D. (1987) *Indictment: Power and Politics in the Construction Industry*. Faber and Faber, London.

[12] Wilson, I. (1978) Scenarios. In: *Handbook of Future Research*, J. Fowles (ed.), pp. 225–248. Greenwood Press, California.

[13] Technology Foresight (1995) *Progress through Partnership, Vol. 2. Construction Office of Science & Technology*. The Stationery Office, London.

[14] Van der Heijden, K. (1996) *Scenarios – the Art of Strategic Conversation*. Wiley, New York.

[15] Zentner, R. (1975) Scenarios in forecasting. *Chemical Engineering News*. October, 22–35.

9 Marketing and promotional strategies in construction

Prepared by Dr C.N. Preece, University of Leeds

Introduction

Underpinning the practice of marketing is a set of ideas broadly defined as a '*marketing philosophy*' which states that a firm needs to identify the needs and wants of its customers and clients and satisfy those needs and wants at a profit [1]. This philosophy on the surface seems very oriented towards a capitalist model of a society. However it is noted that profit can encompass social profits as well financial profits. Marketing (or client) orientation refers to how the whole firm is oriented towards a market place however socially or politically constructed, how all the functions of the business are working to satisfy the client [2]. Marketing strategies are how marketing and promotional objectives are going to be achieved. The process of strategy formulation, analysis, choice and implementation is applied to the marketing processes. As was seen in Chapter 4 firms have increasingly sought to strengthen marketing capabilities as a defence against recession. Marketing has assumed a central place in strategic management of construction firms.

The *marketing function* may be the responsibility of top managers or directors in small firms. As a company becomes larger, marketing departments become distinct functions and may be organised by functions, products (types of project/size of projects), geographical regions or types of customers.

Marketing and promotion are relatively recent developments in the management of construction organisations [3, 4, 5, 6]. Many firms have been found to buy in the services of advertising, market research, corporate identity and promotional design consultants [7]. The limited empirical research has found that there is considerable misunderstanding about what marketing is and how it may be applied. Marketing is not always given the necessary support from senior management in construction organisations and would seem not to be integrated with the other functions of the business, i.e. production, human resources, etc.

This chapter begins by outlining a number of complementary approaches to marketing which have largely been developed in other manufacturing or service sectors, but which may be seen to be appropriate in the construction industry. First, it defines the marketing concept and orientation which demand that the 'philosophy' becomes embedded in the corporate culture of the enterprise. The main approaches reviewed include the concepts of relationship marketing, service quality and customer satisfaction, internal marketing and customer care. Second, the chapter examines the practical issues in the application of marketing strategies including market choice and segmentation, the services marketing mix in construction and promotional strategies in communicating the corporate brand. The final section addresses the need for construction businesses to focus on customer service and the problems to be overcome in applying the approaches developed in the chapter.

The marketing concept

The Chartered Institute of Marketing defines marketing as 'the management process responsible for identifying, anticipating and satisfying customer needs profitably' [8]. A detailed definition of the marketing concept is provided by the Institute of Marketing as the 'management function which organises and directs all those business activities involved in assessing and converting purchasing power into effective demand for a specific product or service and in moving the product or service to the final customer so as to achieve the profit target or other objectives' [8]. Marketing is also defined as '...identifying and satisfying the needs and wants of consumers by providing a market offering to fulfil those needs and wants through exchange processes profitably' [9].

The common theme which is emerging from the above definitions is that they all focus on the management process of establishing customers' requirements for the purpose of satisfying them at a profit to the company. There will be other stakeholders to satisfy, namely employees and customers although modern business seems overly concerned with satisfying shareholders' needs. The starting point of this process is the identification of customers' needs, wants and demands. Those needs, wants and demands should be translated into products and services that fulfil customers' requirements. The guiding concepts for customers to choose between the different offers lie in several factors, which may be regarded as the cost, values provided and their satisfaction with those products and services. The selected products and services need to be delivered via an exchange process for money. This interchange process takes place over a period of time and transactions, during which a kind of relationship may be established.

Marketing orientation and relationship marketing philosophy

Marketing orientation is defined as the organisation-wide intelligence generation, dissemination and response to current and future customer needs and preferences. This indicates that marketing-oriented concepts focus their efforts on a number of factors. These include market focus, customer focus and co-ordinated marketing.

Comparing marketing and selling concepts indicates that marketing focuses on customers' needs, while selling focuses on the sellers' needs. A selling-oriented organisation starts with manufacturing their products, then focuses on the existing products and tries to sell them via hard selling and promotion to generate profits. A marketing-oriented organisation defines its target market carefully, concentrates on customer needs, co-ordinates its activities to achieve customer satisfaction which in turn generates the profits. Thus, the selling-oriented organisations take the 'inside-out' perspective while the marketing-oriented organisations take the 'outside-in' perspective [1].

Customer orientation requires the organisation to put the customers at the centre of their activities [9]. A customer-oriented organisation would stay close to its customers and track their satisfaction over the time. Customer feedback needs to be communicated to all personnel in order for them to be aware of their contributions to achieve customer satisfaction [9]. Customer complaints and quality problems need to be discussed at all levels of the organisation to find ways of improving the company's systems or processes.

Co-ordinated marketing indicates first, that all marketing divisions, sales force, advertising, marketing researchers and so on should be co-ordinated from the customer perspective. Second, marketing departments should be co-ordinated with all the other departments within the organisation. Companies should focus internally, on their internal systems and employees, as well as externally, on their customers. However, there is no point in promising high service quality before the company's employees are ready to deliver the promised service. Companies' employees should be able and willing to serve their customers; they must be well selected, well trained and motivated to do their jobs.

The relationship marketing philosophy has been defined as 'all marketing activities directed towards establishing, developing and maintaining successful relational exchanges' [10]. Considerable emphasis is placed on generating relationships with customers and throughout the supply chain based on high degrees of trust and commitment. Blois [11] describes the long term nature of this approach and the need for 'mutuality of interest' between all the parties. Relationship marketing requires genuine commitment and focus on continuous improvement [12, 13].

Service quality and customer satisfaction

Marketing orientation which incorporates market intelligence, enables companies to focus their quality programmes directly on customers' needs and requirements and assures that market intelligence is the central guidance of the service quality and customer satisfaction strategies. Development of concepts like 'service quality', 'customer service', 'customer focus and orientation', 'customer satisfaction', 'customer retention and loyalty' etc. fit well the basic objectives that marketers hold for the marketing and marketing orientation. Effective implementation of these concepts should lead to a sustainable competitive advantage, increase customer loyalty and in turn enhance the ultimate business performance and profitability of the company. Companies' marketing orientation culture creates the sense to all employees that they have a role to play in providing excellent service quality that achieves customer total satisfaction and loyalty.

Internal marketing

Work by Albrecht & Zemke [14] and Rosenbluth & Peters [15] has found that the more successful service companies are those which believe that employee relations are reflected in customer relations. Management carry out *internal marketing* and create an environment of employee support and reward for good service performance. Rosenbluth & Peters [15] go so far as to say that the company's employees, not the company's customers, have to be made number one if the company hopes to truly satisfy its customers.

A market orientated corporate culture and its impact on increasing customer satisfaction and employee commitment may be measured through examination of a number of areas [16, 17]. For example, the degree to which top managers within a business are orientated towards the market, risk aversion, inter-departmental conflict and reward system orientation. Beckham [18] and Parasuraman *et al.* [19] assert that levels of customer satisfaction are a function of the difference between perceived performance and expectations. Total quality is the key to value creation and companies need to become customer satisfiers and advocates, focused on whole processes, not just on functional marketing.

The work of Crosby [20], Deming [21] and Hiltrop [22] encourages Western companies to learn from the Japanese, such as Toyota, and adopt the total quality management (TQM) philosophy. This philosophy stresses that quality should not be managed just in terms of the interface between customers and suppliers, but encompass all relationships within the organisation through the creation of internal marketing programmes [23]. Hiltrop [22] examines the just-in-time philosophy which places considerable emphasis on increasing flexibility of the workforce and high employee involvement. Emphasis needs to be on the internal dynamics of the company. Meeting the requirements of the internal customer is as important as meeting the needs of the external customer [24].

There is also a need to measure and track customer satisfaction and dissatisfaction by making it easier for the customer to complain. Research has found that 95% of dissatisfied customers do not complain, but just stop buying [25].

The challenge is to create a company culture such that everyone within the company aims to 'delight the customer'. The work of Schneider & Bowen [26, 27] and Parkington & Schneider [28] within banking services has examined relationships between employee and customer perceptions of service climate. These studies have identified that customer attitudes about the quality of service were strongly related to employee views of the service customers received. Also, customers often equate services with the employees who render them and there is a general blurring of the boundaries between employees and customers.

Service businesses are more difficult to manage using a *traditional marketing* approach [29]. In manufacturing, the product is fairly standardised and sits on a shelf waiting for the customer. Gronroos [30] has argued that services marketing requires a combination of external, internal and interactive marketing. Internal marketing essentially describes the training and motivation of management and employees in serving customers well [31, 32]. Interactive marketing describes the employees' skill in serving the client [30, 33].

One of the major ways to differentiate a service firm is to deliver consistently higher-quality service than competitors [34]. Customers compare the *perceived service* with the *expected service* [35]. If the perceived service falls below the expected service, customers lose interest in the provider. If the perceived service meets or exceeds their expectations, they are more likely to use the provider again [32].

Internal customer satisfaction

It is very important to be able to answer the question: 'who is the customer?' Bowen [36] suggests that:

> '...everybody should see himself as a customer of colleagues, receiving products, documents, messages, etc. from them, and he should see himself as a supplier to other internal customers. Only when customers are satisfied has a job been properly executed – it is the satisfied customer that counts irrespective of whether he is external or internal'.

It is only after identifying the internal customer that internal marketing can be applied and internal customer satisfaction can be achieved.

Customer care

Cook [37] considers customer care to be about management of 'the total consumer experience of dealing with the producer.' This involves the producer in controlling and managing 'customer confidence', from the moment they are aware of the product or service to the point where they become part of it. Customer care is about 'managing perceptions as well as realities'. Daniels [38] asserts that these perceptions need to become the 'base-point' from which to make improvements to the service provided.

Clutterbuck [39] asserts that customer care is a fundamental approach to standards of service quality, covering every aspect of a company's operations from design, packaging, delivery and service. Client care initiatives need to permeate every part and activity of an organisation [40]. It involves a complex series of relationships between customers, individual employees and the organisation. It is a means of 'establishing customer-supportive attitudes and behaviour' [41]. Customer or client care can be defined as the identification and management of critical incidents in which customers come into contact with the organisation and form their impressions of its quality and service. The organisation's aim is to provide customer satisfaction [42].

Wellemin [43] introduces the variety of tangible and intangible elements of customer care. Tangible elements include physical features of a product, i.e. its size, colour etc. Intangible elements are more difficult to define and are related very much to the service provided. For example making the customer feel secure, trusting and well disposed towards the supplier and individual members of staff.

Smith & Lewis [44] support these views and consider customer care to be a 'philosophy' of treating customers and clients well and keeping them informed. They also introduce the notion that implementation of the customer care philosophy is dependent on a change in the way that employees are cared for by the company, in terms of management style and working conditions.

From these broad definitions of customer of client care it may be seen that company initiatives need to integrate not only product and service quality, but also marketing and personnel practices if they are to be successfully implemented. In essence, customer care is not only about improving systems and procedures but needs to become a guiding philosophy, part of the shared values, culture and mission of the company.

The concept of customer care has been widely used in manufacturing, service and public sector industries. Blackman & Stephens [45] recognise that the use of customer or client care initiatives in the public sector is on the increase due to the introduction of market forces. A paper by the Audit Commission [46] argued that customer care would be essential if local authorities were to become 'competitive' and questioned whether councils viewed the public as customers with 'views and choices' which should be used in policy planning and implementation. The whole privatisation of public utilities has created a new emphasis on the customer. Major parts of the National Health Service (NHS) have undertaken quality and customer care programmes intended to re-orientate traditionally hierarchical and paternalistic-based cultures into commercial enterprises [37].

According to Clutterbuck & Kernaghan [41], 'sooner or later, effective total quality management programmes run into the problem that most customer complaints are to do with the quality of service'. Customer care may be regarded as the next phase in the evolution towards a customer-oriented company.

Marketing strategies – market choice and segmentation

Firms normally cannot service all the customers in a mass market. The firm has to identify a segment within which to compete. Target marketing enables a firm to identify marketing *opportunities* more effectively.

- *Consumer* markets consist of purchasers and/or individuals in their households who intend to consume or benefit from the purchased products and who do not buy products for the main purpose of making a profit, i.e. a house buyer.
- *Industrial* markets (like construction) consist of individuals, groups or organisations who purchase the services of the industry in order to produce their products or services, i.e. McDonalds, Barclays Bank, ICI, etc.

Market research

In the formulation of marketing strategies, the firm needs an *intelligence system* in order to monitor and analyse the external marketing environment. Market research is the systematic collection, recording and analysing of data concerning the products and services offered and customers and clients. Up-to-date information gained from an effective marketing research audit assists in reducing the risk associated with marketing decisions. Market research can take many forms including desk research of newspapers, business press, construction trade press, planning leads and *networking*; or the use of personal contacts in the industry to identify future clients and projects.

Marketing research for a construction contractor may gather information on:

- demand for certain types of projects; buildings or structure
- categories of client, identifying specific needs and wants
- identifying suitable target market segments
- monitoring the marketing and promotional activities of competitors.

Customers in construction

The firm may identify three basic types of customer:

- *Key clients* – those customers who place repeat orders with the firm and whose loss would have a significant effect on the firm;
- *Existing non-key clients* – those of the firm's customers who approach them infrequently;
- *Non-clients* – customers for whom the firm has not yet worked.

Key client marketing is important in construction where contractors are wanting to obtain more repeat orders and negotiated contracts. It requires relationship building between the construction firm and the client organisation, negotiation skills, internal co-ordination, back-up services, after-sales care and attention and improved market research.

For effective marketing and promotional strategies, the construction firm needs to identify *client buying behaviour*. This involves understanding the process by which a client decides to buy the services of a contractor and those who influence the client in this decision.

In recent years, attempts have been made to develop customer care initiatives in construction [47, 48]. Customer care is the identification, management and control of customers' confidence from the moment they come into contact with the organisation to beyond the point they become part of it. It involves a complex series of relationships between customers, individual employees and the organisation, covering every aspect of the organisation's operations in order to improve the quality of service to the customer.

Who is the client in construction?

The clients in construction are not 'uniform' or 'average' organisations. The objectives of one client organisation may be quite different from those of another. Turner [49] considers clients across five categories: property and development companies, investors, occupiers, local and central government authorities and quangos. Each type of client has different priorities, needs and expectations of the construction service and will require different approaches. An aim of client care must be to add value to the client and their business.

In addition to the 'end-user', the construction client team includes many different types of consultants who offer design and cost control services. It is important to establish their 'buying behaviour'. The involvement of the client team provides for a much greater involvement in the service than in other service industries.

Contractors need to know how their services, marketing efforts, managers and staff, facilities and brand image affect clients' and their advisors' perspectives. They also need to know how the competition are perceived. The firm in practice needs to establish why the client continues to negotiate and place contracts with them. Is this merely repeat business and just how loyal is the client? This information needs to be incorporated into the client care programme.

The only way to find out whether a company has met, or indeed exceeded client expectations, is by asking them. This should identify strengths and weaknesses in the service experienced, and will aid in improving processes in the future.

Who is on the 'front-line'?

Essentially anyone who comes into contact with the client and/or their advisors may be considered to be in the 'front-line'. Contact with the client commences with early meetings, telephone conversations, interviews and presentations during the pre-qualification phase of a project and continues through negotiations and tendering procedures, the award of the contract, construction stage, completion of the building or structure and during the after-care phase. The 'front line' staff in contracting will change throughout the different stages. Unlike other industries, 'client-contact' staff will be at almost all levels in the company, from director level down to labourers on site.

The front-line is drawn from a variety of different functions and departments of the business which need to be co-ordinated and to view each other as internal customers. The client care initiative will need to be driven from director level down if the required level of change to attitudes and behaviour is to be achieved, and if an 'internal marketing' culture is to be established. The client care initiative would aim to improve the quality of services provided by the project and site managers and others during their interface throughout these processes. It would involve identifying the needs of clients during these phases and where improvements were necessary in the service provided by front-line and support staff.

The marketing mix in service industries

The marketing mix consists of five decision areas:

(1) *Product* – Services are intangible and are difficult to standardise. They are people intensive. The marketer needs to identify the features of the service and how these provide benefits to the client. These are essentially concerned with price, time and quality [50, 51].
(2) *Price* – Services are difficult to price before the service has been completed.
(3) *Promotion* – Unlike tangible goods such as cars or washing machines, construction services cannot be displayed. The features and benefits of the service need to be demonstrated through promotional activity. Promotion makes the intangible, tangible, by creating an image for the customer of the service he will receive.
(4) *Place* – Where the service is performed for the customer. For example a bank or travel agents in the high street. In construction, the client's office and the project site.
(5) *People* – All the people of the firm need to be oriented towards the market place and the client. Many managers and staff throughout the firm will have some contact with the client or their representatives on site, in the office, or over the telephone. The *network* of people involved in the *marketing exchange* includes the contractor, subcontractors, suppliers, architects, quantity surveyors and engineers. The success of the project and the quality of the overall service depends on how these individuals, groups and organisations interact. The construction industry has a tradition of an adversarial culture which has hindered successful implementation of the marketing philosophy.

Promotional strategies

Promotional objectives include [52, 53].

- increasing market share, winning more contracts from the same or similar clients;

- informing the market place of the existence of the firm;
- providing information on how the service of the firm is unique and different form other competitors in the market place;
- encouraging a favourable business climate to encourage repeat orders;
- launching a new product or service, i.e. design and build or management contacting to the market;
- developing and promoting the corporate brand.

The firm seeks to *communicate* with potential customers and *persuade* them to buy the firm's products and/or service. The firm may use a combination or *promotional mix* of efforts (see Fig. 9.1).

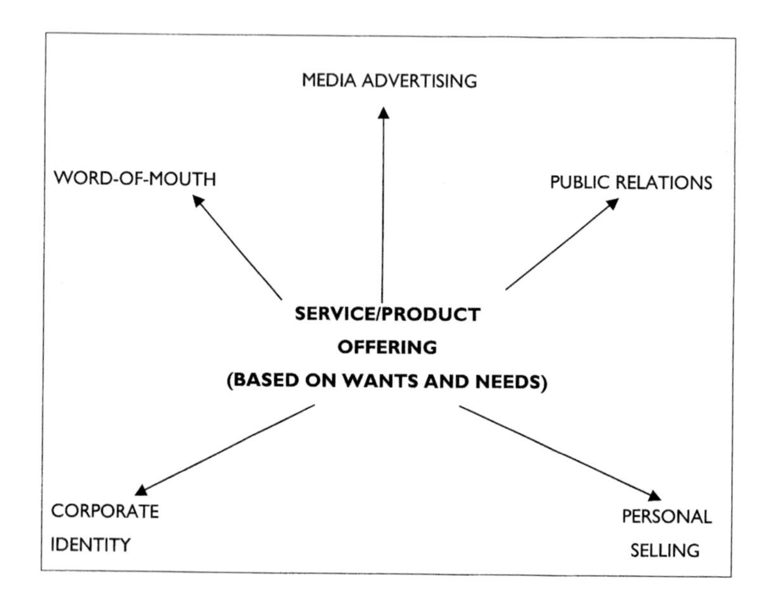

Fig. 9.1 Promotional mix.

Promotional media (personal/non personal)

Personal selling and presentation

The most persuasive promotional media available to the contractor are personal selling or more specifically pre-selection/qualification interviews and team presentations. Client organisations and those who influence the client, i.e. architects, quantity surveyors and others are interested to know how the firm is going to undertake the proposed project to time, cost and to the required level of quality [54]. The firm will need to present those members of the construction team who are likely to be involved with the project rather than pure salespeople.

Non personal media

Existing media includes advertising in print and broadcast forms, i.e. newspapers, magazines, television, commercial radio, etc. Public and press relations activities gain editorial coverage for the firm's activities and achievements. This requires good relations with appropriate local, national, specialist newspapers and journals and broadcast media.

Created media includes direct mail, targeted to specific segments and audiences. Success depends on selection and accuracy of mailing lists and the design of promotional brochures and other material. Publicity material such as brochures need to be designed to stress the benefits of the service to the client. Publicity activities can involve opening ceremonies, sponsorship of sporting events or art exhibitions, visits by VIPs to project sites, etc.

Audiences

Clients of the industry are in the public and private sectors and range from individual house buyers through to multinational corporations and foreign governments. Through market research, firms need to establish their clients' needs and wants in relation to the proposed project and what they value in the services they receive. The firm also needs to establish how the client is influenced to be able to decide on the most effective media. Word-of-mouth and editorial comment on a firm's image, its products and services have been found to be more influential than press advertising, particularly in relation to contractual services. The firm needs to establish what the various members of the client's decision-making unit are looking for and to promote these through the appropriate media.

Messages

Promotional messages will vary according to the products or services provided by the firm. For example a design and build contractor will stress expertise and reputation in design, in addition to management of the project. Messages should be designed to communicate the features of the service and the benefits of these to the client and the proposed project.

Messages conveyed in selling contractual services need to stress [55, 56]

- general experience and reputation
- financial standing and record
- quality assurance registration
- a satisfactory prior business relationship with the client.

Criteria for *qualifying* – being shortlisted to submit tenders – have been found to include:

- reputation for completing projects on time
- recent experience of similar projects
- the expertise of the management team.

The primary factors favouring the winning of a contract have been found to be:

(1) lowest contract price
(2) reputation for time performance
(3) expertise of project management team
(4) personal contact.

The factor leading to a contractor *losing* a contract is mainly a high contract price.

Co-focusing on customer service – the problems to be overcome

In today's competitive climate, marketing customer service is an effective tool to enhance the firm's position in the market place [57]. It should be seen as the principal way to create and sustain a competitive advantage and build a strong and lasting relationship with clients. It may be the vehicle for construction businesses to differentiate themselves from their competitors.

More demanding and discerning clients are increasingly looking for high quality of service from largely technical project management teams. This service must be experienced from initial contact, throughout the phases of a project, through completion and handover and beyond. Achieving a competitive advantage through contractual service marketing is increasingly requiring closer understanding and identification with the client, internal cross-functional partnering and active executive level support.

Services marketing requires special adaptation in construction due to the unique characteristics of the industry. It is one of the most diverse businesses in the world and projects are generally unique. Furthermore, it covers a very wide range of products and the people working in it come from a broad range of professional backgrounds and crafts.

The construction industry is diverse and often involves a variety of parties with different and conflicting interests. It is the variety of interest that provides the fertile environment for conflict in the industry. Moreover, this problem will escalate with the traditional separation of design and construction. Furthermore, there are other factors which increase client dissatisfaction:

- The finished product does not meet the agreed specifications.
- The companies focus on the technical aspects of the product rather than elements of the service.
- Conflicts may arise between the personalities and management of the companies with those of the client which increase client dissatisfaction.

All the above factors will undoubtedly have an effect on the relationship between clients and their contractors and sub-contractors. Therefore construction firms should recognise the impact of product quality and quality service on this relationship and in turn on client satisfaction. Whilst research has found that construction companies appreciate the need for good staff-to-client interfaces during all stages of the construction project, there is little evidence that comprehensive care programmes are being implemented. Some companies have commenced training of staff in telephone behaviour and limited use of training packages has been identified. There is a need for better co-ordination and co-operation across the different departments or functions in contracting to improve the overall service delivered to clients. A number of firms have appointed 'customer care' managers and have allocated single points of contact for particular types of client.

Companies need to establish the priorities sought by the client. A project of high quality, handed over to the client, tells only half the story. The standard of service in providing the project may have been less than satisfactory, with major conflicts between the parties rather that a relationship marketing culture focusing on satisfying the clients' needs and wants.

Firms need to know how their products and services, people, facilities and brand image affect clients and their advisors. They also need to know how the competition are perceived. The firm in practice needs to establish why the client continues to negotiate and place contracts with them. What are their priorities? Is this merely repeat business, and are they continuing to look for better service from the competition? Just how loyal is the client?

Contractors and consultants need to be able to manage word-of-mouth, previous clients' experiences and use them as independent testimony to the quality of the service provided. Clients will use such information along with their own experiences to benchmark against competitors. The only way to find out whether a company has met, or indeed exceeded client expectations, is by asking them. This should identify strengths and weaknesses in the service experienced, and will aid in improving processes in the future.

There is little evidence to suggest that pro-active market research is conducted to identify opportunities and client requirements. Marketing would not seem to be seen a part of strategic level planning. According to the limited research into corporate image building, there are few data that construction companies are effectively differentiating themselves from their competition, particularly during the pre-qualification process [58, 59]. Client teams have indicated that there is a lack of attention to the tailoring of marketing approaches to their specific needs, and a lack of professionalism in presentations. This would suggest a deficiency in understanding of the basic principles of marketing communications.

Construction company cultures have been described as *production*, or at best, *sales* orientated. Where firms have been found to have developed marketing/sales departments, these are not integrated with the other parts of the business or a catalyst for corporate cultural re-focusing onto clients' needs.

Contracting in construction demonstrates all the characteristics of industrial services generally. Marketing approaches, as adopted by product manufacturers or consumer services are largely inappropriate. A new paradigm which emphasises development of an integrated external client focus, internal service culture and interactive marketing skills of construction personnel would seem to be appropriate for contractors.

Research identifying clients' needs and expectations reinforces the arguments that word-of-mouth recommendation and personal experiences of service providers are far more important than any promotion used by construction firms [60]. Differentiation in contracting is therefore more to do with the people engaged in the process.

Many of the reported reasons for the sense of dissatisfaction with construction are to do with elements of the service, rather than the more technical aspects of the end-product, the building or structure. Despite industry-wide efforts such as the reports by Latham [61] and more recently Egan [62], which attempted to encourage adoption of more partnering in procurement, construction is typified by a highly adversarial climate between the various parties to the contract. This is frustrating the development of more effective marketing in construction.

References

[1] Kotler, P. & Armstrong, G. (1993) *Marketing: an Introduction*, 3rd edn. Prentice-Hall, NJ.

[2] Morgan, R.E. & Morgan, N.A. (1991) An exploratory study of market orientation in the UK consulting engineering profession. *International J. of Advertising*, **10**: 34–48.

[3] Fisher, N. (1986) *Marketing in the Construction Industry; a Practical Handbook for Consultants, Contractors and Other Professionals*. Longman Group Ltd, Harlow.

[4] Pearce, P. (1992) *Construction Marketing: a Professional Approach*. Thomas Telford, London.

[5] Fellows, R. & Langford, D. (1993) *Marketing and the Construction Client*, Chartered Institute of Building. Bourne Press, UK.

[6] Preece, C.N. & Barnard, L. (1999) *Marketing Strategies of Engineering Consultancies Given an Increasingly Competitive Environment*. Proceedings of the 4th National Construction Marketing Conference, 1st July 1999, pp. 14–24. Oxford Brookes University.

[7] Preece, C.N. & Male, S.P. (1997) *Promotional Literature for Competitive Advantage in UK Construction Firms*, J. of Construction Management & Economics, pp. 29–42. E & F N Spon, London.

[8] Institute of Marketing (1973) *Marketing in the UK Construction Industry*. HMSO, London.

[9] Woodruffe, H. (1995). *Services Marketing*. Pearson Professional Ltd, London.

[10] Morgan, R.E. & Hunt, S.D. (1994) The commitment–trust theory of relationship marketing. *J. of Marketing*, **58**, No. 3.

[11] Blois, K. (1995) Relationship Marketing in Organisational Markets - its Information Needs. Templeton College of Management Research Paper No. 95/6, Templeton College, Oxford.
[12] Christopher, M., Payne, A., Ballantyre, D. (1991) *Relationship Marketing*. Butterworth-Heinemann, Oxford.
[13] Landeros, R., Reck, R., Plank, R.E. (1995) Maintaining buyer–supplier relationships. *International Journal of Purchasing and Materials Management*, **31**, No. 3.
[14] Albrecht, K. & Zemke, R. (1985) *Service America!* Dow-Jones-Irwin, Homewood, IL.
[15] Rosenbluth, H.F. & Peters, D.M. (1992) *The Customer Comes Second*. William Morrow and Co., New York.
[16] Kohli, A.K. & Jaworski, B.J. (1990) Market orientation: the construct research propositions and managerial implications. *Journal of Marketing*, **54**, 48–60.
[17] Jaworski, B.J. & Kohli, A.K. (1993) Market orientation: antecedents and consequences. *Journal of Marketing*, **57**: 53–70.
[18] Beckham, J.D. (1992) Expect the Unexpected in Health Care Marketing Future, *The Academy Bulletin*, **12**: 39–56.
[19] Parasuraman, A., Zeithaml, V.A., Berry, L.L. (1985) A Conceptual Model of Service Quality and its Implications for Future Research, *Journal of Marketing*.
[20] Crosby, P.B. (1979) *Quality in Free*, McGraw-Hill, New York.
[21] Deming, W.E. (1982) *Out of Crisis*, MIT Press, Cambridge, MA.
[22] Hiltrop, J.M. (1992) Just-in-time manufacturing: implications for the management of human resources. *European Management Journal*, **10**, No. 1: 37–43.
[23] Chaston, I. (1994) Internal customer management and service gaps within the UK manufacturing sector. *International Journal of Operation and Production Management*, **9**: 68–82.
[24] Atkinson, P.E. (1990) *Creating Culture Change: the Key to Successful Total Quality Management*. IFS Publications, Bedford.
[25] Technical Assistance Research Programs (TARP) (1986) *Complaint Handling in America*. US Office of Consumer Affairs Study.
[26] Schneider, N. & Bowen, D.E. (1984) New services design, development and implementation and the employee. In: *Developing new services*, W.R. George & C.E. Marshall (eds), pp. 92–101. American Marketing Association, Chicago, IL.
[27] Schneider, N. & Bowen, D.E. (1985) Employee and customer perceptions of service in banks: Replication and Extension. *Journal of Applied Psychology*, **70**, No. 3: 423–433.
[28] Parkington, J.J. & Schneider, B. (1979) Some correlates of experienced job stress: a boundary role study. *Academy of Management Journal*, **22**: 132–149.
[29] Kotler, P. & Bloom, P.N. (1994) *Marketing Professional Services*. Prentice-Hall, NJ.
[30] Gronroos, C. (1984) A service quality model and its marketing implications. *European Journal of Marketing*, **18**, No. 4: 160–179.
[31] Berry, L. (1986) Big ideas in service marketing. *Journal of Customer Marketing*, Spring, Editorial, pp. 3–6.
[32] Heskett, J.L., Sasser, W.E., Hart, C.W.L. (1990) *Service Breakthroughs*. Free Press, New York.
[33] Mills, P.K. (1985) *Managing Service Industries: Organisational Practices in a Post-Industrial Economy*. Ballinger, Cambridge, MA.
[34] Baron, S. & Harris, K. (1995) *Marketing: Text and Cases*. Macmillan Press, London.

[35] Berry, L. & Parasuraman, A. (1991) *Marketing Services Competing Through Quality*. The Free Press, New York.

[36] Bowen, D.E. (1986) Customers as Human Resources in Service Organisations. *Human Resource Management*, **25**, No. 3: 31–39.

[37] Cook, S. (1992) *Customer Care: Implementing Total Quality in Today's Service-Driven Organisation*. Kogan Page Limited, London.

[38] Daniels, S. (1993) Customer Care programmes. *Work Study*, **42**, No. 1, 74–86. MCB University Press Limited, Bradford.

[39] Clutterbuck, D. (1988) 'Developing Customer Care Training Programmes', Industrial and Commercial Training, Vol. 20, No. 1, MCB University Press Limited, Bradford.

[40] Bee, F. & Bee, R. (1995) *Customer Care*. Institute of Personnel and Development, London.

[41] Clutterbuck, D. & Kernaghan, S. (1991) *Making Customers Count: A Guide to Excellence in Customer Care*, pp. 62–74. Mercury Books, London.

[42] Thomas, M. 91987) Customer care: the ultimate marketing tool. In: *Reviewing Effective Research and Good Practice in Marketing*, R. Wensley (ed.). Marketing Education Group, Warwick.

[43] Wellemin, J. (1995) *Successful Customer Care in a Week*. Institute of Management, London.

[44] Smith, A.M. & Lewis, B.R. (1989) Customer Care in Financial Service Organisations. *International Journal of Bank Marketing*, **7**, No. 5: 13–22.

[45] Blackman, T. & Stephens, C. (1993) The internal market in local government: an evaluation of the impact of customer care. *Public Money and Management*, **13**, No. 4: 84–98.

[46] Audit Commission (1988) *The Competitive Council*. HMSO, London.

[47] Preece, C.N. & Shafiei, M. (1998) *Development of Client Care Initiatives in Construction Contracting Organisations*, pp. 586–596. Association of Researchers in Construction Management (ARCOM) 14th Annual Conference and Annual General Meeting, University of Reading.

[48] Preece, C.N. & Shafiei, M. (1998) *Client Care in the Construction Industry – a Competitive Marketing Tool*, pp. 52–56. Proceedings of the 3rd National Construction Marketing Conference, Oxford Brookes University.

[49] Turner, A. (1997) *Building Procurement*, 2nd edn. Macmillan Press Ltd, Basingstoke.

[50] Preece, C.N. & Tarawneh, S. (1997) *Developing High Quality Design and Build Services: Closing the Four Managerial Gaps*, pp. 71–77. The 2nd National Construction Marketing Conference, Oxford Brookes University.

[51] Preece, C.N. & Tarawneh, S.A. (1996) *Re-Orientating the Construction Team to Achieve Service Quality for Client Satisfaction*, pp. 37–42. The 1st National Construction Marketing Conference, Oxford Brookes University.

[52] Nickels, W.G. (1984) *Marketing Communication and Promotion*, 3rd edn. Grid Publications, Columbus, Ohio.

[53] Rossiter, J.R. & Percy, L.P. (1987) *Advertising and Promotion Management*. McGraw-Hill, New York.

[54] Preece, C.N. & Tarawneh, S. (1997) *Developing High Quality Design and Build Services: Closing the Four Managerial Gaps*, pp. 71–77. The 2nd National Construction Marketing Conference, Oxford Brookes University.

[55] Preece, C.N., Moodley, K., Habeeb, M. (1995) *The Management of Direct Selling and Pre-qualification Team Presentations for Competitive Advantage in Contractual Services*, pp. 11–20. Liverpool John Moores University Conference on Practice Management for Land, Construction and Property Professionals, Merseyside Maritime Museum.

[56] Preece, C.N., Male, S., Moodley, K. (1996) *Targeting, Tailoring and Trimming – the More Effective Use of Promotional Literature in the Marketing of Contractual Services*, pp. 103–111. The 1st National Construction Marketing Conference, Oxford Brookes University.

[57] Baron, S. & Harris, K. (1995) *Service Marketing, Text and Cases*. Macmillan Press Ltd, London.

[58] Preece, C.N., Putsman, A., Walker, K. (1996) *Satisfying the Client Through a More Effective Marketing Approach in Contracting*, pp. 5–9. The 1st National Construction Marketing Conference, Oxford Brookes University.

[59] Preece, C.N., Shafiei, M.W.M., Moodley, K. (1997) *The Effectiveness of Promotion in the Private House Building Industry*, pp. 78–83. The 2nd National Construction Marketing Conference, Oxford Brookes University.

[60] Preece, C.N., Putsman, A., Walker, K. (1996) *Satisfying the Client Through a More Effective Marketing Approach in Contracting*. 1st National Construction Marketing Conference Proceedings. 4th July 1996. pp. 5–9. Oxford Brookes University.

[61] Latham, M. (1994) *Constructing the Team*. HMSO, London.

[62] Egan, J. (1998) *Rethinking Construction*. Department of the Environment, Transport and the Regions. The Stationery Office, London.

D Summary

10 A synthesis of strategic management in construction

Introduction

The construction industry is a large industry consisting of small firms. The contracting side of the industry is staffed by operatives who are young, male and casually employed. There is a strong craft-based tradition, either on-site or with off-site manufacture. The industry comprises consultant service firms and construction firms that can be viewed as either service providers or manufacturers. The 'professions' can be distinguished by gradations of specialism – the design and cost specialisms associated with the built product and those associated with the process, such as project management, value management and programming. The firms that populate the industry, regardless of their size, are economic as well as social entities and have the capacity to adapt to and learn from environmental changes, provided the staff within those firms recognise the need to change and can do something about it.

The industry can be characterised by a series of overlapping markets in terms of size, geographic location, type and complexity of project. Within this market structure, demand is dominated by the private sector, coupled with the increasingly speculative nature of demand. Chapter 2 indicated that the industry now faces a paradigm shift, with a move towards lower value, shorter duration projects and a distinct geographic imbalance of workload activity dominated by the South East, with pockets of activity in other parts of the country. An increase in the number of smaller projects has consequences for the numbers of managerial, administrative and clerical grades within construction firms, with a consequential impact on overheads, profit and human resource policies. Earlier chapters have argued that the general level of economic activity affects profitability in the industry, with long-term changes in demand evidenced as sectoral shifts in market structure. With long-term shifts in markets, especially the increase in repair and maintenance work, the move towards smaller project sizes in some market sectors will necessitate different work patterns, skills and training requirements.

The industry is characterised by a wide diversity of clients, local, regional, national and international, with differing degrees of knowledge of the industry, who range in size from individuals to multinational enterprises and from local to national governments. Clients are seeking new ways of interfacing with the industry through a diversity of services.

This sets the context for the final chapter, which brings together the earlier themes of the book and synthesises and integrates them to raise implications for strategic management for the construction industry.

Strategic management in the construction industry

The construction industry environment has experienced a series of evolutionary changes since the Second World War, characterised as operational, strategic and competitive change. Firms in the industry have had to adapt to these in different ways. Adaptation occurs through senior managers making choices amongst alternative courses of action and structuring the organisation in such a way that it 'fits' with the new environmental conditions that it faces.

Environmental change in the industry can be classified into two major and three contingent types:

- Recurrent change, which is incremental and permits an organisational memory to be developed and requires no major change in the relationship or fit between the firm and its environment. Operational change identified above is a form of recurrent change, where the managerial focus is on process improvements and internal efficiencies.
- Transformational change is fundamental and shifts the relationship between the firm and its environment. Strategic and competitive change fall within the domain of transformational change. Strategic change, which is discontinuous, fundamental and radical, is likely to be felt right the way through the organisational structure, from the strategic apex to the operating core. Competitive change, whilst fundamental in the medium to long term is incremental. It creates a sustained and deep-seated pressure on a firm. An example would be the continued shift of demand from the public to private sectors. Competitive change, combining strategic and operational change, suggests facets of environmental discontinuity tempered by a more evolutionary shift. In essence, this type of change requires a simultaneous focus on corporate, business and production strategies. Transformational change, since it involves a fundamental shift in the relationship of the firm with its environment requires a decision-making strategy of 'vigilance'.

Within contracting and the professions there has been a restructuring of firm sizes predominantly towards the larger or smaller firms. The natural consequence is that medium-sized firms remain under increasing competitive pressure from firms at both ends of the scale.

The commercial environment of consultants has changed dramatically. They now face competition on price, their roles are being redefined through a range of different procurement routes and forms of contract, they are now able to take up directorships in companies and there have also been changes in the size of practice, with involuntary specialisation through involvement with particular clients and the larger firms becoming much more involved in the strategic end of projects. Liberalisation of the consultant service sector, coupled with the dramatic changes in demand highlighted above has also meant that consultants have experienced strategic change.

Industries and markets in construction

A firm requires inputs and uses them to create outputs to sell to customers to meet demand. The concepts of market and industry are distinct when considering a firm. In economics, a market is a demand side concept and is buyer (client) generated. Five competitive forces determine the underlying economic structure of an industry. Porter [1] sees these forces being shaped by:

- The relative power of buyers;
- The relative power of suppliers;
- The level of inter-firm competition;
- The ability of substitute products or services to perform the same function as that demanded by customers;
- The potential for new entrants to enter the industry.

An analysis of these five forces provides a long-term structural view of the impact of the industry on a firm or firms. Industries tend to be relatively stable and are generated by the outputs from production capabilities. It is recognised that whilst Porter's competitive advantage school of thinking and the formal strategic planning ethos have dominated thinking, there are alternatives. The power culture school sees strategy being formulated by political horse-trading by senior managers. This power broking influences strategy and the adopted strategy has to be consistent with the existing sets of power interests. Equally emergent strategy may be an alternative formulation appropriate for construction. Emergent strategy is a pattern of discussions and actions taken over time coming together to form a strand of strategic thinking. This incrementalist model may be made by making small opportunistic steps. This is mirrored in small day to day steps such that construction-to-contract, procurement routes and tendering strategies set up the market structures for firms in the industry, twelve types of procurement route and two types of tendering have been identities and there are significant differences in their use between the civil and building engineering industries. The full gambit of procurement routes is used in the latter, whereas the former has a much narrower focus on the use of procurement routes, with a greater emphasis on traditional procurement.

There are two primary market types operating in the construction industry, namely, construction-to-contract or contracting and speculative work. There are different economic forces at play in each. Each market type requires a different set of managerial and commercial skills.

Depending on the procurement route adopted, construction-to-contract has institutionalised distribution channels operating for contractors' services through clients' consultant advisers. In construction-to-contract, price is determined prior to production, the reverse of traditional manufacturing. In speculative work construction companies are involved in demand creation for their products and price is determined after production. This has a closer affinity to manufacturing industries.

The variability in demand shapes the nature of the construction industry and the firms operating in it. The industry is not a single industry but is made up of a number of different market sectors defined by:

- Geography
- Size of project by value
- Complexity – both technical and managerial
- Type of project.

Aggregate demand at the industry level comprises the demand profiles for each of the market sectors. Sectoral demand may vary significantly compared with overall aggregate demand. For the strategist, indicators exist that permit long-term thinking, such as underlying client needs, the value of new orders (an indicator based on contractor returns and subject to short term variability and value of construction output), the aggregate of all construction work across the public and private sectors. The latter is the most frequently used indicator and is relatively stable over time. The construction industry appears no worse than traditional manufacturing industries in having to cope with wide variability in demand. In this sense it is not unique.

Sectoral shifts in demand have significant implications for market specialisation and the ability of firms to respond to changing demand profiles. There are also geographic imbalances in demand, with the Southeast dominating market trends. It is, however, more sensitive to economic cycles than other parts of the country. The market is also dominated by short duration, low value projects. Depending on market sectors, 56 to 90% of projects are less than £2m with a significant proportion less than 12 months' duration. A further trend identified in earlier chapters has been an increased emphasis on speculative development requiring increased investment in land banks and hence capital lock-up. This has reduced the financial flexibility of firms. Demand changes and high levels of market turbulence have impacted medium-sized firms in economically depressed areas of the country the most.

Strategists need a framework for understanding the business environment. This can best be thought of as the general and task environments facing the firm as set out in Chapter 2.

Entry and exit barriers exist in an industry

Entry barriers prevent firms entering an industry or market sector, exit barriers prevent firms from leaving an industry or market sector. Construction has always been viewed as having low entry and exit barriers. Porter identified six different types of entry barrier. An analysis in an earlier chapter indicated that each was shown to operate in construction in different ways. In construction-to-contract, entry barriers are subtle. They involve consultant advisors acting as intermediaries between client and contractor and the type, size, technology and locality of a construction project also act as entry barriers.

Entry and exit barriers that have been explored in construction include:

- Product differentiation in terms of either the service-product or end-product.
- Capital requirements. Construction has always been viewed as requiring low capital requirements. However, research evidence reported in earlier chapters indicates that this is no longer the case for some firms, who require capital to invest in fixed assets depending on their corporate configuration, diversification strategies, for equity investment in projects under PFI schemes or to invest in the development of human assets.
- Switching costs for buyers – the clients of the industry. This will encompass project size, complexity, duration, specialist skills and knowledge requirements.
- Distribution channels, both tangible and intangible. The intangible distribution channels in construction are client advisors for contracting firms.
- Scale economies and the experience curve. This will depend on corporate configuration and the degree of permanency associated with project teams.
- Government policy, a key issue in international activities for construction firms.

Earlier chapters concluded that in the lower value and shorter duration end of the project spectrum entry and exit barriers would be lower. However, for complex projects entry and exit barriers are likely to be high, requiring service-product differentiation combining reputation, expertise, financial engineering and price to drive key competencies and managerial and technical know-how. In addition, diversified construction firms create different entry and exit barriers in the industry. Strategic groups comprise firms that follow similar strategies and entry and exit barriers become transposed into mobility barriers between groups that prevent firms moving easily between groups.

Construction has also been called 'fragmented' and this is related to the geographically dispersed project-based nature of the industry that requires manufacture to take place at discrete, multiple locations. Fragmentation also relates to the split between the new build and repair and maintenance sectors. New build markets have a range of project sizes, from the very small to the very large. Project type and location have an impact on the extent to which the industry is, in reality, fragmented. Large, complex, high value contracts require managerial expertise, technical know-how and financial stability as prerequisites. The result is that only a relatively small number of contractors can tender for such projects. Pre-qualification procedures ensure that the numbers of firms entering the market are small. The concept of the contestable market deals with market relationships for these types of projects. In the repair and maintenance sector projects have a greater tendency to be small and local firms have an advantage. Location also plays an important part. The geographically dispersed nature of demand may favour the local contracting community with their own local, or regional client and consultant contacts.

The nature of the firm

A firm is a social organisation that has profit as one of its potentially many objectives or goals. A firm can also be construed as comprising a series of contracts and relationships that add value to customers through delivering quality and offering something distinctive. Some economists have argued that organisations exist because it is more cost effective to handle transactions within a firm than through a series of numerous, ongoing transactions due to the cost of organising and doing business within and between markets over time [2].

The firm has an organisational structure whose primary purpose is to reduce the variability in human behaviour, achieve a common purpose and to co-ordinate tasks allotted by managers to different types of labour. Power is also exercised within the structure and information flows take place between its different constituents. Co-ordination within the structure can take place through five mechanisms: through mutual adjustment, involving considerable informal communication and negotiation, through direct supervision and finally, through different forms of standardisation, namely, through work processes, where the content is specified; through skills, where the training requirements are specified and through outputs, where the results are specified. There are three basic building blocks to organisations' structure. Power exists along a continuum, where centralisation describes the extent to which power is centralised within a few hands or decentralised throughout the structure. Formalisation describes the extent to which codes and norms are explicit, not just in terms of being written down but commonly understood throughout the organisation. Finally, complexity comprises three elements, a horizontal element dealing with the degree of specialisation of tasks, a vertical element dealing with the number of levels in the hierarchy and a spatial element dealing with the extent of geographic dispersion of the organisation.

The organisational structure also involves people structuring their expectations into different situational roles. A situational role will be impacted by the formal and informal structures of the organisation, its technical system and the individuals themselves performing the role. The impact of the technical system and technology are probably the most pervasive in an organisation.

The firm's product in construction

The product life cycle is a key strategic management concept that relates the outputs of a firm to its markets. There are four stages to the life cycle: the development stage, where only a few firms are in the market; the growth stage where the market is exploited by firms; the maturity stage where the market has stabilised and the decline stage, where there are over capacity and too many producers. In construction there are two types of product. The service-product goes through the stages of the product life cycle; the various procurement strategies are mechanisms to select the appropriate service-product to deliver the end product as a completed facility. The end product will go through a project life cycle of concept, design, construction/production, handover and use. From the customer's perspective, they are more concerned with the appropriate service-product to delivery of an end product that is fit-for-purpose.

Core business, core competencies and distinctive capabilities

The core business of a firm has been described as comprising its distinctive capabilities engendered from three primary sources, organisational architecture, reputation and innovation. Organisational architecture stems from the social and commercial relationships that the firm has and it comprises three components, internal relationships between employees, external relationships with suppliers and customers and networks of firms involved in related activities. One of the crucial elements to organisational architecture is relational contacts founded on trust and expectations. These are the moral and psychological contracts that the firm has, which may or may not be underpinned by a legal contract. As an asset, organisational architecture can create and sustain organisational knowledge and routines to respond to change. To be effective as a source of competitive advantage, organisational architecture has to be distinctive to the firm. Reputation, as a competitive asset, has to convey meaningful information to customers. It is built through a process of continued success, the converse of which is that it decays easily through poor performance. Reputation embodies the long-term experience of the firm and creates the likelihood of repeat business. It also forms part of the value creation process, permitting the firm to charge a premium provided it exceeds the cost of maintaining and upgrading it. Innovation, as a competitive asset, is difficult to secure rewards and a profit from on its own but when linked with organisational architecture, in particular, can embed a process of continuous innovation in the firm.

Core business for construction firms was seen as encompassing some or all of the following:

- Where the firm had a longstanding interest and had built up considerable expertise
- Substantial turnover

- Profitable or expected to be so
- Reasonable market growth or a captive market
- Low capital requirements.

Firms who identified it as contracting ($N = 10$); of which three identified this as purely building for their firm, six identified this as building and civil engineering and one identified it as civil and process engineering.

When translated into strategic areas, the following emerged in sample research conducted by Hillebrandt *et al.* [3]:

- Two firms who identified their core as contracting and housing
- Two firms who identified their core as contracting, housing and property
- One firm who identified its core as contracting, housing and minerals
- Two firms who identified their core as contracting and mining (and other)
- One firm who identified its core purely as housing.

Core competencies are the collective learning of the firm across its skill, production and technology bases and provide value to the customer. There is a difference between threshold competencies and core competencies, which go beyond the former. Using terms from earlier chapters, threshold competencies are the qualifying levels of service that are required to survive and remain as serious competitors, and would include competency to undertake the work in question. This would be based on expertise and reputation. The determining level of service is that which is unique to the seller, adds value to the customer and provides a competitive advantage. These are the basis of core competencies. Core competencies are likely to be implicit but need to be expounded explicitly for competitive advantage to be clear, they should also be difficult for competitors to imitate. They are likely to be rare, complex and embedded in organisational knowledge and practice. They are not about delivering the product.

Design, procurement and construction were seen as core construction industry functions by the eLSEwise research project [4], with the contractor's prime function seen as mobilising the specific resources to construct a unique product in a given time. These are not core competencies but qualifying levels of service or threshold competencies.

Innovation in construction

Two main forms of innovation have been identified, radical shifts, which are rare in construction, and incremental innovation, which is common in construction, last for many years and may go unnoticed. Incremental innovation will normally follow radical shifts. Porter identifies innovation as one of the key issues in sustained competitive advantage, although Kay [5] has indicated the need for that to be linked to organisational architecture to be viable. A number of themes have been highlighted in earlier chapters concerning innovation in the construction industry. Four distinct types of innovation have been identified:

(1) Technological innovation, which utilises new knowledge or techniques to provide a product or service at lower cost or higher quality.
(2) Organisational innovation, which involves 'social technology' and not the technical system. This type of innovation is about changing the relationship between behaviours, attitudes and values of people in the firm. In construction a number of examples have been identified by Lansley [6], for example, new types of business organisation, new forms of contract and procurement and the opening up of new markets. Partnering, Prime Contracting and PFI are the most recent forms of organisational innovation.
(3) Product innovation, which involves advances in technology resulting in superior products or services.
(4) Process innovations, which concern increasing efficiencies but may not involve significant advances in technology.

Much of the innovation in the industry occurs at the workface with individual craftsmen on special projects. The high levels of sub-contracting in the industry mean that this type of incremental innovation is also available to competitors in the market place. Innovation in equipment or materials lies outside the industry and hence construction is more involved with innovation diffusion. The effective utilisation of equipment and materials by construction firms creates knowledge based competitive advantages. Again, high levels of sub-contracted plant hire and leasing means this is potentially available to competitors. Lansley [6] has identified the importance of the impact of long waves in the economy for innovation in construction firms. His empirical data suggest that different types of innovation may be more appropriate in periods of decline than in periods of recovery. During the 1960s companies were involved in process innovation, focusing on operational change and a task-oriented style of management. In the 1970s when strategic change impacted firms, product innovation came to the fore, with a management style that was people and corporate oriented. During periods of competitive change a different type of innovation is required, one that combines both product and process innovation, the latter focusing on gains inefficiency whilst the former concerns investing in and launching new competitive initiatives to gain a longer term advantage over competitors.

The next section reviews issues surrounding typologies of firms.

Typologies of firms

Mintzberg [7] identified five types of firms:

- The simple structure, more applicable to small firms
- The machine bureaucracy, which co-ordinates by standardising work processes

- The professional bureaucracy, which co-ordinates by standardising skills, uses convergent thinking and pigeon holes solutions to client problems.
- The adhocracy that uses mutual adjustment as a co-ordination mechanism. There are two types of adhocracy, the operative adhocracy which is client focused, comes up with innovative solutions through flexible teams and blurs the distinction between operational and administrative tasks and the administrative adhocracy
- The divisional structure that co-ordinates by standardising outputs.

In addition, Miles & Snow [8] have tackled the issue of organisational typologies from the perspective of managers' perceptions, the internal power and political structures of organisations and the relationship between strategy, structure and process. They identified four types of firms:

- The defender, which sticks to what it is good at
- The prospector which sacrifices internal efficiency for innovative ways of offering new products or services
- The analyser, which uses an internally stable and efficient part of the organisation to feed an innovative part of the organisation
- The reactor, which does not have a consistent set of internal relationships that fit easily with the demands of the environment. In Porter's terms it is 'stuck-in-the-middle' and is neither one thing nor the other.

Other organisational typologies identified included Ansoff's [9] proactive systematic, proactive adhoc and reactive types of organisations; the sector structure, the holding company, the subsidiary, the matrix organisation that is either permanent or temporary, the multinational corporation and the virtual enterprise.

A number of these typologies have emerged from the endeavours of academics to classify organisations from research data, others have emerged as responses to the industrial environment.

Strategy

There is a considerable amount of confusing terminology surrounding strategy. It is normally defined as the means to meet ends, where the means are a set of rules to guide organisational decision-makers throughout the organisation and the ends are a series of measurable objectives. Strategy can be either explicit or implicitly understood as a shared collective belief in the direction that the organisation is to take. Some writers have also seen strategy in terms of a resource allocation process within defined business scope, whilst others have seen it as a series of outcomes that affect the goals of an organisation and its stakeholders. Due to the confusion surrounding the term strategy other writers have also taken the term back to its roots in the military, seeing strategy as being about locating a firm's operations or competitive arena, in an industry, industry segment or geographic location, and tactics, the mechanisms to achieve a strategy, by using particular organisational functions or parts of the organisation as resources and know-how. The military perspective, advocated by Vasconcellos [10] but also discussed by other writers such as Mintzberg *et al.* [11], has much to commend it for construction, given the importance of the locational issues surrounding the production base. In the military context, strategy and tactics *both* have a short and a long-term perspective attached to them, rather than the normal discussions surrounding strategy, which see it as important decisions that have long-term ramifications for the firm. The military perspective would see strategists within construction, for example, as having a short-term strategy of competing on a particular project type in a particular location to test the water. This may then lead to a longer-term strategy to exploit that market sector and project type by setting up a permanent office in the locality. The tactic may be to use the best team to ensure the greatest chance of success and then the longer-term tactic is to use that team to train up others when a more permanent operational office is set up.

Mintzberg [12] has attempted to clarify the debate surrounding strategy and has proposed that there are five aspects to strategy:

(1) That strategy involves planning a course of action - plan
(2) That strategy involves manoeuvring and scheming to outwit competitors - ploy
(3) That strategy can be discerned over time when looking at a firm's consistency of action and behaviour - pattern
(4) That strategy is also concerned with making decisions about locating a firm in its environment - positioning
(5) That strategy can also involve a deep-rooted consensus amongst the members of a firm on managers' intentions over time - perspective.

Mintzberg et al. [11] have also seen strategy going from deliberate to mostly emergent. The military perspective would encompass all five of the above, with strategy and tactics covering points 1 and 2 and strategy covering points 3, 4, and 5.

A hierarchy of strategies has been identified in earlier chapters:

- Corporate strategy is holistic and concerned with the activities of the whole company across its organisational configuration of different business units.
- Business strategy is concerned with competing in different markets or industries.
- Operational or functional strategy is concerned with the operating core or functional departments. A focus on on-site production and the associated processes would be an example in construction.

The diversified construction firm uses the full hierarchy of strategies and will face different strategic time horizons. In contracting a period of two years is perhaps the maximum time horizon for strategists since this tends to be the maximum length of many construction projects. With the large firms, who can obtain some very large projects this can be extended to three years. In property development the planning horizon can be four to five years and for manufacturing subsidiaries the planning horizon can extend into the long term. At main board level these different time horizons will have to be integrated into an overall corporate plan.

Structuring the strategic management process

Hillebrandt *et al.* [3] put forward a useful two-phase corporate planning process stemming from the types of environmental change that firms face. The planning process comprises two activities, seeking internal efficiencies and/or responding flexibly to the external environment. The flexibility phase concerns formulating questions about the firm's business and identifying alternative paths for its development. The efficiency phase follows on and is about deciding on a course of action and developing the most efficient way of implementing it. Short-term budgets are an efficiency issue whereas focusing on using resources innovatively to adapt to the business environment is a flexibility response.

During periods when the external business environment is relatively stable, firms can focus their attention on preparing detailed plans for achieving high levels of internal efficiency. However, during periods of considerable environmental turbulence firms need to attend to flexibility issues and use resources in new ways. During such periods focusing on detailed planning for internal efficiency is counterproductive when the need is to concentrate more importantly on flexibility and the demands of the external environment.

Three components comprise the strategic management process:

- Strategic formulation is about analysing the organisation's current situation and where it wants to be in the future. A series of tools and techniques can be used to assist the strategist and fed into the decision-making process. The outcome of the formulation process is a range of alternative strategies and tactics that can be used. A trio of components of the strategic management process can be seen. Each has a number of alternative courses of action.
  - Generic strategies of:
    - cost leadership
    - differentiation
  - Directional strategies:
    - Opt for operational change
    - Opt to consolidate or stabilise
    - Retrench through divestment

- Penetrate existing markets, either with new products in existing markets or new markets with existing products
- Diversify:
  - Within the broad confines of the industry and associated industries – related diversification
  - Outside the industry, products or markets totally – unrelated or conglomerate diversification
  - Internationalise – go outside the current domestic markets into overseas markets and spread the geographic base of the firm
  - Integrate forward, which takes the firm closer to the customer and concern securing greater control over outputs
  - Integrating backwards which gives the firm control over inputs.

- Mechanisms:
  - Internal growth by using the firm's capital to develop its own business activities or set up separate business activities under a corporate configuration
  - External growth by using its capital to secure other businesses. This can be achieved either through acquisition, a more hostile activity or through merger, a more collaborative approach
  - A combination of the two through some form of 'grand strategy' [2]
  - Strategic choice is about making decisions from a range of alternatives that will determine the future strategic position of the firm in its industry and within which strategic group, where it decides to compete against its competitors, which client needs it intends to meet and how it intends to position itself against the five competitive forces that determine industry structure. The criteria for making strategic choices are the degree of acceptability within the organisation, the suitability of the choices made when compared against the intended strategy and the environment faced by the firm and the feasibility of implementing the chosen strategy
  - Strategy implementation occurs through 'organisation' and requires appropriate feedback mechanisms in place to ensure that the strategy as implemented is consistent with the strategy as conceived, evolved and decided upon. Implementation forms part of the tactical response of the firm to strategy. Implementation requires three fundamental questions to be asked:
    - Who is responsible for strategy implementation?
    - What must be done for successful implementation?
    - How will implementation work?

There are seven primary mechanisms for implementation – a tactical response:

(1) Plans and policies at corporate, SBU, operational and function levels depending on the organisational configuration
(2) A budgetary framework for resource allocation

(3) Reward systems
(4) Political systems
(5) Control and integration systems that encompass the structure in general, the hierarchy, teams and team management, rules and procedures
(6) Training and development systems
(7) Feedback mechanisms in the form of key performance indicators and measurable objectives.

Construction firms demonstrating good strategic management processes [3]

- Formulate an overall strategy at the strategic apex based on intuition and informed awareness
- Expect operating units to develop and present their own plans to the main board to consolidate into a single plan
- Use planning departments to provide contextual background information, undertake analysis and develop the board's thinking into operating plans
- Have mechanisms that permit strategies to be changed if the external and internal circumstances necessitate it
- Adopt a combination of top-down and bottom-up, that is, the loose–tight approach. The senior managers at main board level provide the goals and vision, whilst those at divisional and regional levels provide the detail and identify the opportunities and actions that are consistent with plans at main board level. The planning horizon is typically three to five years or longer. The main area where firms fall down is in the implementation process for plans. Regular reviews of progress also assist with the monitoring and implementation process.

The next section reviews the decision-making role of the strategist.

The decision-making role of the strategist

The strategic management process involves managers making decisions about the firm and its relationship with the external business environment – its degree of 'fit'. Within this context decision-making modes have been identified. Strategic decisions are about positioning the firm in its environment, they are non-programmed decisions that are made often in conditions of highly ambiguous and uncertain information and are made by senior managers at the strategic apex of the firm. Administrative decisions concern resource allocations and decisions about organisational structuring, whilst operating decisions are involved in the transformation process of the firm, the primary value-adding activities involving inputs and outputs.

The role of strategists, as senior managers in a firm, is to manage the firm going from where it is now to where it wants to be in the future using the strategic management process. This will involve strategists in making choices from amongst a series of potentially viable options. The strategists' primary role is to manage the firm through strategic and competitive change. Operational change is more likely to be managed lower down the managerial hierarchy depending on the size and structure of the organisation. There are five decision-making modes, within two broad bands of inappropriate and appropriate ways of handling change:

(1) Inappropriate actions:
  (a) Misperceiving change triggers and remaining with the status quo when this is unsuitable
  (b) Implementing incremental change when a more radical response is required
  (c) Correctly perceiving the requirements for change but avoiding taking action due to a belief that there is no solution.
  (d) Correctly perceiving the requirements for change but believing here is insufficient time for implementing a response.
(2) Appropriate action – correctly perceiving change and the necessary response.

Six procedures have been identified to assist with the management of change:

(1) Identifying strategic and non-strategic types of decisions and ensuring time and resources are committed to assist the strategic decision-making process.
(2) Avoiding time traps that come from a focus on operational issues or encountering a crisis situation that may result in an inappropriate response.
(3) Using a filtering system to target the appropriate response at the right level in the firm.
(4) Using a buffering system to allow strategists time to focus attention on strategic decisions and change.
(5) Using champions of strategic change whose primary role is to draw the organisation's attention to the change management process and what is being implemented.
(6) Allotting and monitoring time allocated to strategic and non-strategic activities.

Flexible firms that are successful at change management demonstrate the following characteristics:

- Senior managers are perceived by staff to have a clear future vision of where the firm is going and are committed to clearly defined and stated objectives for achieving well-defined market goals.
- Senior managers are concerned about the welfare of staff.

- An organisational climate exemplified by an internal consistency between the management style used by senior managers and that preferred by staff.
- High levels of staff morale and job satisfaction.
- A history of effective change management.
- Effective market intelligence systems.
- Effective corporate planning systems.

Inappropriate change management activities mirror the above and include:

- A lack of concern with human resource management or a corporate perspective.
- A lack of attention to people and a high level of organisational politics leading to low staff morale and lack of effectiveness.
- A lack of understanding of how to undertake environmental scanning, particularly among medium-sized firms.
- A misunderstanding of corporate planning and budgetary control systems, seeing them as synonymous when they are not.

The elements of competitive strategy

The elements of competitive strategy that need to be considered by strategists are:

- The internal factors within the firm, its strengths and weaknesses and the key values of the strategists.
- External factors to the firm, the industry opportunities and threats and the expectations of society about firms and the nature of business in general.
- The scope of the business, those customers served and their needs - the strategic square of Vasconcellos [10] or the SBA of Ansoff.
- Resource utilisation, the areas of distinctive competence and the knowledge and skills that comprise these.
- Areas of synergy, the interactions of activities and distinctive competencies across businesses or services.
- Value activities that stem from structure and process, managing people, the technical systems used by the firm and the linkages in the value chain between the transformation process, supplier inputs and outputs demanded by customers. Value activities and the associated value chain are a product of the firm's history and its strategic management process.
- Sources of competitive advantage, where the firm has superiority over competitors, often located in the technical core.

Sources of such an advantage require constant improvement and upgrading. They can accrue from a number of distinct sources within a hierarchy of advantages. High order sources are:

- proprietary process technology
- product differentiation by offering a unique product or service
- advanced skills and capabilities through specialised and highly trained personnel that come from either an internal technical capability or having close working relationships with leading customers – relational contracts
- brand reputation, which represents the cumulative marketing efforts and customer relationships that are locked into a company through high switching costs for other firms
- sustained cumulative investment to create tangible assets or intangible assets that go towards creating reputation, customer relationships and specialised knowledge.

Low order sources that may be easily replicable by competitors are:

- Low labour costs
- Low materials costs

These form a common basis in the contracting business.

Design is a high order knowledge based advantage. Contractors are in the business of creating knowledge-based advantages through:

- Financial engineering
- Alternative ways of organising the production process
- Using technology in different ways during the production process and across teams, as technological innovators
- Creating new services
- Creating new forms of organisational structure
- Using skill and craft innovation, which occurs at the workface but is impacted considerably by a firm's sub-contracting strategy.

The marketing function and its contribution to competitive advantage

Marketing is a management function closely bound up with the strategic management process [13]. Its focus is on identifying, analysing and anticipating the needs of customers and supplying them at a profit within the capabilities of the company. Marketing is linked to the demand side concept of the market. The following comprise the marketing philosophy:

- A customer orientation. Customers can be classified as:
  - Key customers or clients. These are important for repeat orders and negotiated work. They have a potentially significant impact if their business was lost.
  - Existing non-key customers or clients. These are customers who approach the firm infrequently.

  - Non-customers, these are clients who the firm has not yet worked for.
- A total company effort.
- A profit objective.

The marketing approach segments the market place in terms of customer needs and the products or services bought or used. Segmentation creates target markets requiring the following to be ascertained:

- market growth potential
- dominant competitors
- entry barriers
- components of value added.

Chapter 9 identified differences between market-oriented and selling-oriented firms. Market-oriented or customer-oriented firms work 'outside-in' out by defining target markets, understanding customers' needs and then co-ordinating the firm's activities to achieve customer satisfaction. They try to stay close to the customer and use a variety of feedback mechanisms to support this. A selling-oriented firm works 'inside-out' by focusing on the seller's needs, manufacturing products and then trying to sell them. Construction firms have been identified from research as being production-oriented or at best selling-oriented, clearly not customer-oriented. Client dissatisfaction comes not from the technical delivery of the end-product but the components going to make up the service-product. Consultants are involved in service delivery. Contracting can be argued to be either a service process or manufacturing process, or both, and this has already been alluded to in earlier chapters, with the different types of contracting firms that populate the industry.

The marketing mix comprises the five 'Ps':

- Product – which is about providing customer benefits. The project-based nature of contracting means that contracting firms can face substitutability of service delivery of the end-product by other contractors. Pre-qualification makes the assumption that those contractors selected are capable of undertaking the work. In other words their technical capabilities are similar. Depending on the pricing culture in evidence by the client – price can be the final arbiter. Reputation, to an extent, embodies the fact that previous buyers are satisfied with performance in delivering the end-product.

- Price services are difficult to price before completion but a contractor, through the bidding price within the pre-demand markets has to provide such information, as do consultants if they are required to provide a lump sum fee. The price quoted by a contractor represents the cost of providing the structure – the manufacturing element – and the cost of service delivery. Thorpe & McCaffer [14] indicate that in a competitive bidding situation 90% of the tender is made up from the estimated cost of the work and the remaining 10% is the mark-up for risk, overhead and profit. Contractors believe that their competitive advantage comes from the pricing of preliminary items. In a competitive bidding situation where the lowest price may be the deciding factor, quality may suffer during the production phase.

- Promotion is about making explicit the features and benefits. Promotional activity provides a mechanism for differentiation. In contracting and consultancy it can assist in getting on to select lists and to obtain negotiated contracts.
- Place – where the service is performed. For design teams it will be in their offices, in the client's offices, on-site or in the contractor's offices. A structure is erected in a locality where the client has a need and the service is consumed at the point of delivery, the construction site. For contracting firms their 'shop window' is the site.
- People – are those comprising the network making up the marketing exchange. Expertise is a people issue and is a doing skill and hence upgradable. Potentially it requires some form of certification when embodied into reputation. This can be viewed from three perspectives:
  - Certificated professional qualifications, the external validation of skills
  - In-house training courses and on-the-job training
  - Certificated and non-certificated external training courses.

Customer care is about managing the perceptions and the realities of customers when they experience 'product' delivery. It is about controlling and managing customer confidence and concerns managing those critical incidents where the customer comes into direct contact, and experiences, the organisation. Customer care involves both tangible and intangible elements, with the former related to the physical feature of the end-product and the latter related to the service provided. More demanding clients are now looking for a high quality of service from a largely technical team. This can included seeking a closer identification and understanding of the client, active executive support and serviced delivery that is experienced through all phases of the project.

Chapter 9 also pointed out the dangers to firms of not taking on board marketing or undertaking it incorrectly. These dangers included insufficient attention being paid to market research and identifying opportunities or understanding client needs, not seeing marketing as part of the strategic level planning activity, little effort being taken to differentiate from the competition, a lack of tailoring to particular client projects and a lack of professionalism in presentational teams. When summed up these demonstrate a lack of awareness of the principles of marketing as one of the weapons for gaining competitive advantage. Chapter 9 indicated a new paradigm is required, one that encompasses client oriented service-delivery infused throughout the delivering organisation as part of its culture. This will require interactive marketing skills to build relational contracts. Within this new paradigm, identifying client needs and expectations, ensuring a good personal experience of the delivery to generate word-of-mouth recommendations is much more important than other types of promotional activity. An accelerated move away from the adversarial nature of contracting appears the only way forward for contractors.

Tools, techniques and methodologies for competitive advantage

A number of tools, techniques and methodologies have been highlighted and discussed for use in the strategic management process. In summary, these are as follows:

- The SWOT analysis.
  - It has been argued by some that strengths and weaknesses can be external to the firm and difficult to overcome, namely, structural issues rooted in the economics of the industry, the strategic positioning of the firm in the industry and its strategic group. Strengths and weaknesses can also be internal to the firm and can be related to its ability to implement its chosen strategy, which stems from the capabilities of people, its management or from the operating core. Strengths and weaknesses also relate to what the company does well and does this matter in the market place and what does it do poorly and does that matter in the market place?
- Competitor analysis focuses on analysing the actions and moves of competitors in order to understand the context within which the company's competitive strategy has to be developed. There are four major components to competitor analysis, namely, ascertaining their:
  - Future goals of competing firms at corporate and SBU levels
  - Implicit and explicit current strategies
  - The key assumptions held by senior managers about the firm and the industry
  - The key capabilities: distinctive competencies, sources of advantage, growth potential and capacity to handle change, staying power, both within the industry and also with chosen strategies.

Ohmae [15] has questioned the relevance of competitor analysis, arguing that it can waste time and energy and can act as a constraining force on managers' thinking. His argument is that a firm's managers should implement their chosen strategy to meet customers' needs with vigour. It has been argued in an earlier chapter that for contractors, competitor analysis needs to be carefully focused due to the large number of firms that may be potential competitors. It has also been suggested that a firm's 'peer group' may be the best focus of attention – its strategic group.

Strategic group analysis is concerned with identifying those firms that are pursuing the same or similar strategies in an industry. Strategic groups are relatively stable over time and the key influences for identifying them are the degree of interaction between firms in the market place and their extent of target customer overlap. An attempt to apply this form of analysis to construction [16] has highlighted the need to use a methodology suitable for a complex industry.

The 'strategic square' is closely related to the 'strategic business area' (SBA) concept [10]. The strategic square or SBA concept circumvents the issue of industry, market, service-product and end-product classifications. They require the following to be identified:

- a future market need, a demand-oriented focus
- a project technology, defined by the social and technical systems on a project
- a customer with a need
- a geographical location where the customer has that need.

Value chain analysis has both an internal and external component: it is concerned with understanding inbound logistics, the transformation process related to production; and finally, out-bound logistics concerned with delivery of the facility to the client, in this case hand over. The value chain is a product of a company's history; culture; strategic management process; and, the cost and resourcing implications of these. A number of discrete value activities have been identified in the pre-tender process in contracting [14, 17]:

- estimating
- contracts planning/management
- the procurement/buying function
- the sub-contract tender process
- tender adjudication, identified as a strategic and operational activity involving members of the strategic apex and middle line.

In the production stage of contracting other activities that impact the delivery of value to the client include:

- labour strategy - directly employed versus LOSC
- the site management role
- the organisation and co-ordination of sub-contractors
- site recruitment practices
- industrial relations in terms of employment conditions - procedures for bonus and overtime payments
- organisation and control of boundary management between sub contractors.

Portfolio management techniques for reviewing and developing separate business units. These techniques are based on the experience curve and the product life cycle. There are three fundamental characteristics of such techniques:

- The market growth rate
- The relative market share of the firm compared to the market leader
- The revenues that are generated from products or SBU activities.

In construction, portfolio management techniques can be applied at the corporate level, for service-products, end-products and for the management of multi project strategies. Scenario testing permits strategists to create alternative futures, either through economic forecasting, visioning or identifying branching points where discontinuities may occur. Cross impact analysis can also assist scenario testing by looking at the strength of impacting events that may either be unrelated to a situation or enhancing the occurrence of an event. To be worthwhile, however, scenario testing must be credible, useful and understandable by managers.

Construction corporations and contracting corporations

This section brings together a number of the key concepts discussed earlier to describe a possible typology of corporations in construction.

Construction corporations

There is a strong differentiation between constructors. There are those that provide management services and are service oriented. There are those that build the physical product and are manufacturing-oriented. A number of consequences stem from this and may reflect different industrialised responses to deal with risk. The former provide a management service for a fee to co-ordinate an operating core that is sub-contracted out, moving risk on to sub-contractors through contractual conditions. The latter see themselves as managing risk on behalf of their clients and are prepared to take this on board for a price. They are also much more intimately involved in the production process, which increases their capacity to manage risk.

Within construction firms, five business areas can be determined easily:

- Civil engineering
- Building engineering
- Property development
- Estate development (housing)
- Construction product manufacture.

They represent the traditional areas for diversification by construction firms. Earlier chapters have raised a number of issues concerning a typology of construction firms, especially the larger firms. It is not an easy task now to describe the nature of the large construction corporation. First, many firms have diversified to such an extent that contracting is only a small percentage of their overall business. Second, some construction firms have diversified into construction related industries, whilst others have diversified outside of construction related activities. Third, there are also those firms where contracting still remains a significant part of their turnover and fourth, some firms in the industry, viewed traditionally as construction firms, are considering selling their contracting divisions due to low profit levels. A typology [16, 18] that may assist in unravelling this is:

- A contracting corporation is a large firm where contracting remains a significant percentage of turnover, in the order of 95% of sales coming from contracting. This is the single-product firm.
- A construction-dominant corporation is one where some diversification within construction related industries has taken place but this represents a small percentage of the firm's turnover, perhaps in the range 5–30%. Contracting remains part of the core business. This is the dominant-product firm.
- The construction-related corporation, where a firm has diversified into construction related markets but no one product or service accounts for more than 70% of total sales. This is the related-product firm.
- A construction-conglomerate corporation is one that has its origins in the industry and which could take two forms:
  - where diversification has taken the firm outside of construction related industries and where no one product accounts for more than 70% of turnover,

  or
  - where diversification has taken outside of the construction related industries *but* it has divested itself of its contracting business.

The above also relate to the evolutionary stages of construction corporations, where the single-product firms are in the first stages of their evolution, with low organisational complexity and little interest in activities outside their area of specialisation. Dominant-product firms are those firms that have diversified into operations and markets that offer scope for integration with their basic business line. They are in an intermediate phase of development. The related and unrelated firms are those that have evolved into more complex entities, either associated with or including business activities outside their industry of origin. They are in the third and fourth stages of development. A refinement of this nomenclature is the extent of 'relatedness' between diversified activities [19]. Constrained diversification is one where activities are closely related by virtue of requiring similar skills in technology, production, marketing and management know-how. Linked diversified activities are only vaguely associated with each other within the corporate configuration and unrelated diversification has much the same meaning as previously. This latter nomenclature is more related to the underlying logic of a strategy when diversification takes place. Figure 10.1 presents schematically a model of a construction corporation.

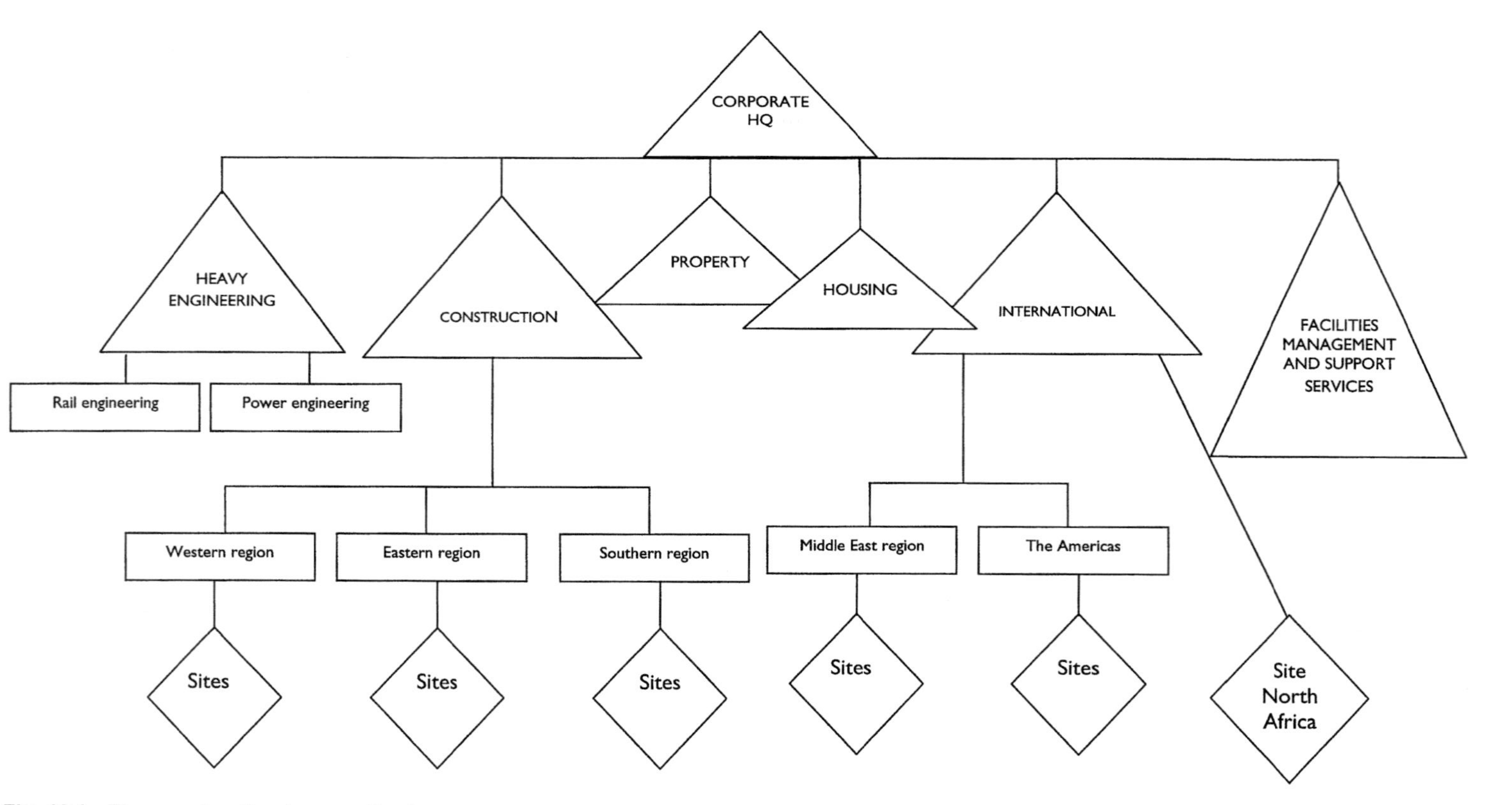

Fig. 10.1 The construction 'corporation'.

Contracting corporations

Contracting firms have been characterised strategically as operating in a merchant–producer role in the industry. This encompasses the supply side merchanting operations, involving trading in goods and services, and the demand side production role encompassing the on-site production process. The contracting firm's organisation balances these two demands through structure and process and uses market power to broker the supply chain to manage value, time, cost and quality. They also use this role to manage workload flexibility to meet the changes in demand and maximise sectoral opportunities. Two distinct merchant–producer types can be isolated. Those firms that are more service-oriented, where the merchanting role dominates and those that are manufacturing-oriented, where the producing role dominates. The other key issue when considering the construction corporation and its contracting operations can become one of the extent to which the diversified firm has the capability within its corporate organisational structure to offer a series of single point offerings to clients, for example, under PFI and Prime Contracting routes.

Structure, strategy and process – managing the diversified construction firm

Organisational structures and roles

Two models of an organisation were presented in Chapter 4. The structural model defines a firm in terms of a strategic apex, middle line, operating core, the techno structure and support staff. The operating core comprises a portfolio of projects.

Potential capacity was used to highlight the knowledge-based capacity of the contractor to gear up to increased workload in this instance usually located at site management level. Due to demand variability contractors have increasingly opted for a structure that is truncated at the operating core through the use of sub-contracting. They have chosen to structure themselves as administrative adhocracies to handle demand variability.

This approach defines the permeable boundaries of the firm and sets the parameters for defining what is internal or external to the firm. The increased use of sub-contracting means that the managerial hierarchy and within contracting firms, have become fragmented [17]. The operating core is characterised by a workforce that is divided by method of payment, employment contract and company loyalty.

The key component of the operating core of the contracting company is management of the project life cycle – the primary task of site management. The operating core of the construction company comprises a diversity of operative attitudes present, including the following:

- A lack of commitment between operatives and employers due to the issue of continuity of work. Many operatives leave projects in anticipation of redundancy; others wish the project to continue as long as possible, which suggests an implicit tactic of slow working by directly employed workers, while a minority see prompt completion as important for the employer's reputation. This analysis suggests employers' interests are best served by piecework-based incentive systems under conditions of operative self-employment in order to maximise productive potential.
- Bonus and overtime chasing is seen as acceptable behaviour by operatives and stems, in part, from a perceived lack of commitment by employers to permanent employment. This suggests that bonus and overtime payments are purely a mechanism for attracting and keeping skilled labour and are not used as a primary method of motivation by management.
- There is disillusionment with the wage structure in the industry. This forces a move towards self-employment which is also compounded by the fact that directly-employed operatives view themselves as becoming increasingly marginalised and ill-rewarded for the level of skill possessed. Structural casualisation characterises the operations of the UK construction industry and to some extent it benefits both employers and employed to the detriment of unionism. However, there are considerable disadvantages - a lack of training and career opportunities, high labour mobility, divisive industrial relations, a continued impact on the image and status of the industry for new recruits. As projects become more complex this is also placing greater burdens on site management.

Construction strategies

Construction firms will, in the future, need strong capital asset bases, even for contracting, due to:

- The need to convince clients of their financial strength
- The increasing use of bonding on projects
- The requirement to put equity into certain projects
- The need to raise money on the stock market using a high asset base.

They have had to explore new ways of obtaining work to maintain turnover related to their core business. They need to capitalise on their special expertise and be much more proactive in marketing. This has required them to differentiate their services. This has been achieved through broadening their existing services, offering new ways of packaging existing services, offering new services or taking over unfinished contracts from bankrupt firms in preference to acquiring that business.

Design and build has increased significantly in importance as a service, whilst the forms of contractor-led management procurement have declined in importance this represents a change in core products. Firms have moved downstream in the project life cycle to provide equipment and furnishings fitouts, structured maintenance, and also facility management. Contractors have moved upstream in the project life cycle into the pre-construction stage by offering a more integrated design and construction service. They have also become more involved in putting together financial packages and identifying potential projects, of particular importance in overseas markets. Financing has become important in the domestic market with PFIs. In addition, some firms have developed specialist expertise in the businesses of their clients, often requiring specialist staff to demonstrate a total capability for delivering the project.

There have now been major changes in firms' strategies:

- A focus back to core business
- Increasing attention paid to overseas markets neglected during the boom period prior to the last recession
- An increasing emphasis on financial management, profits and cashflows
- Increased attention to marketing as a business tool
- Tightening up the structure and organisation of firms
- A continued policy of reducing permanent employment that also covers managerial staff
- A shrinking of training and support for education.

Firms are also aspiring to remain or become large national or international contractors that need to compete against large mainland European contractors. Some firms are clearly setting their sights on becoming major global competitors. The next sub-sections review some of the options available to contractors and the reasons behind diversification.

Diversification in construction has occurred because of the need to [3]:

(1) increase profitable growth
(2) seek different activities in which profitable growth could be achieved
(3) increase efficiency through control of supplies or link activities because they provided greater synergy
(4) use positive cashflow and increase fixed assets
(5) avoid construction cycles and particular clients and markets.

Some diversification strategies that had taken place in the 1980s were undertaken for the right reasons whilst others were opportunistic, in areas where contractors had no expertise and as a result were poorly managed. Diversification strategies included moving into:

- property
- housing

- building materials
- coal and other mining activities
- plant
- mechanical and electrical engineering
- builders' merchants
- construction businesses internationally.

Non-construction related activities included

- time share accommodation
- health care
- airports
- waste disposal.

When recession hit, construction firms abandoned diversification strategies. Following recession, firms have been undertaking diversification again but in a different form.

Business unit strategy: managing divisionalised, regional structures

Regionalisation is the setting up of geographically dispersed operating units or strategic business units (SBUs). The main drivers for regionalisation are growth and expansion. A regionalised structure can also involve expansion and retrenchment, sometimes simultaneously. This will depend on how long the regional structure has been set up and the market conditions it faces within different regions. A regionalised structure brings managers in closer contact with the market place, it gains better access to important production and organisational inputs and creates a better working environment for staff at all levels. It can also be a process to decentralise decision-making. A central theme in creating and managing a regionalised structure within a contracting division is one of managing a process of continuing organisational change and adaptation. Managing a regional structure can involve the setting up or winding down of a series of ongoing regional operating units. It also involves decisions on centralisation and decentralisation of decision-making within regional structures.

The degree of centralisation–decentralisation from the centre has important implications for regionalised structures. If the centre retains a centralised approach to regional structures, the strategists' focus at the centre is likely to be on the similarity of issues and markets between regions rather than on their differences. Decentralisation provides regional senior managers with a great deal more decision-making autonomy and allows them to make greater use of local market knowledge. Centralisation and decentralisation within regional structures also require thinking about senior management succession and associated management development and training.

Strategies at the operating core in contracting firms

High levels of sub-contracting have meant that the main contractor's primary role has now become one of organising, co-ordinating and procuring inputs into the production process; providing services of management expertise, experience, backup and resources from an established organisation and an ability to carry contractual risks and obligations for large and complex projects. The merchant-produced role was alluded to earlier. The on-site production process in construction, unlike manufacturing, is characterised by few routine procedures and is labour intensive. The five major inputs into the production process are:

(1) materials
(2) labour
(3) site management
(4) plant and equipment, which can be owned, leased or hired
(5) finance for working capital.

Project portfolios and potential capacity

The dimensions forming the basis of project portfolios as a concept are:

- Technological complexity
- Project size, normally measured by value
- Type of project, for example, office blocks or industrial units
- Geographic location.

Conflicting views of how contractors use and perceive a portfolio of projects is evident. A market sector view argues that they provide benefits to the larger rather than smaller contractor by:

- Facilitating the use of differential rates of profit between market sectors.
- Minimising the risk of one contract failing.
- Providing greater bargaining power with clients due to other work in hand.
- Providing gains from acquisition of other contractors in terms of access to non-construction assets, rapid market entry without fear of retaliation, additional management expertise, portfolios of contracts and contacts and membership of select tendering lists.

On the other hand, empirical evidence indicates companies do not consider demand in terms of market sectors, with the exception of housing development, but in terms of technologies to execute project types. Managers assess projects in terms of project size, project complexity, construction method and the associated organisational and managerial requirements of the project.

In contracting the production base is transient, one-off and variable. The operating core of a contracting firm comprises numerous production bases, located in different geographical locations, at different stages in their production cycles and with different resourcing requirements depending on the project life cycle operating at that location. Potential capacity stems from the firm's organisational structure and the accumulated knowledge of management and support functions. Potential capacity is the ability to gear up for a higher workload in the operating core.

Sub-contracting as a production strategy within project portfolios

Much of the on-site production process is now sub-contracted out to other firms and is has arisen due to increasing technical complexity of projects, changes in employment legislation over the last 20 years, increasing pressures on employers to reduce fixed costs and the short-term variability of workloads. The result has been a requirement to retain flexibility in the on-site production process. Sub-contracting also provides access to specialist knowledge that could be expensive to retain in-house. It is also a low-cost method of organising the work since parts of the production process are sub-let for a known price through competition.

To be effective, subcontracting requires considerable co-operation between site management and sub-contractors but contractual responsibilities are a continual source of divisiveness and also profitability. Sub-contracting also has considerable social costs attached to it, for example, the apportionment of responsibility for health and safety, training and the undermining of the apprenticeship system. Due to the diversity of inputs, the one-off nature of the product from site to site, the disruptive effects of the weather and the diversity of the workforce, site managers make many impromptu decisions, often without reference to senior managers. This requires a great deal of personal knowledge and an ability to be responsive to a wide range of problem situations. Site management is also the focus of much of the inconsistency between company policy and actual site practice.

Labour only sub-contracting (LOSC) will often attempt to determine from main contractors, in advance, which site agent will be in charge of a project and price their tender submission accordingly [20]. This clearly suggests sub-contractors place a pricing premium or penalty on the abilities of site management.

One of the major defining characteristics of the construction labour force is the differing skill levels. Skill is a social construct and confers different degrees of status, earning power, industrial influence, exclusion and task competency. At the operative level most skills and abilities are readily transferable.

General labourers form the largest group of operatives within construction and account for 25–30% of the workforce in building and over 40% in civil engineering. There is also a marked difference between the skilled labour required for building and civil engineering projects. Of the 70% of skilled operatives in the industry, the four trades of carpenters and joiners, painters, bricklayers and mechanical plant operators account for over half the workforce. Plant operators are employed to a greater extent in civil engineering projects. On large construction sites in particular there is a wide diversity of crafts, often represented by different unions and with differing employer–union agreements in existence.

There are two key issues to consider in procuring subcontract labour as a resource input [1]. The first relates to its scarcity. The more scarce a particular type of labour the greater its bargaining power. The second relates to the extent of labour organisation. Organised labour, through unions or trade associations, has greater bargaining power. Over time, the labour force in construction has become highly fragmented. LOSC has been an emotive subject in the industry for many years and has caused a crisis in trade unionism in construction. Hillebrandt *et al.* [3] conclude that it is here to stay and is deeply embedded in the employment practices of the industry.

In addition to the foregoing, labour mobility and operative attitudes are also influencing factors in the operating core. Labour mobility into, out of and within the construction industry is a key issue in the production process and has implications for skills development in the industry. High levels of labour turnover and stability can coexist easily within construction [21]. The operating core also comprises different mixes of resource inputs, required at different stages of the project life cycle on-site, different types of subcontractor with different trade union memberships and operative attitudes. These are key inputs from the contractor's supply chain into the production process. Site management is, therefore, a key knowledge-based source of competitive advantage capable of building and losing reputation and also includes the capability to manage and use site based innovations that occur at the workface. Site management level is one of the key sources of distinctive capability for competitive advantage in terms of firms' organisational architecture for managing the on-site innovation process and building project site-based reputation for delivery to the client.

Management as a source of competitive advantage

To formulate a competitive strategy it is important to isolate areas where the company has distinctive capabilities. Construction is a knowledge-based industry where similar technologies are available to competitors but it is how it is used that provides the competitive edge. Knowledge resides at regional subsidiary level in terms of detailed market knowledge and client contacts. Knowledge also resides with specialist sub-contractors and again this will be available to competitors through the sub-contracting process. Through relational contracts and the use of key supplier lists, contractors are in a better position to retain preferential access to such knowledge. Earlier chapters identified the fact that there is a hierarchy of sources of competitive advantage and that contractors compete on both low and high order factors, some of which are sustainable only in the short term whilst others have the potential to be sustainable in the longer term. Competitive advantage also requires the analysis of value-adding activities and how these are configured into a value chain. Much of the value-adding activities in construction revolve around team and individual knowledge. Construction relies extensively on project team working and problem solving and isolating competitive advantages in construction can be difficult since there is a high incidence of knowledge-based advantages residing in project teams at different levels in the firm's structure. Some authors have suggested that distinctive capabilities, which provide sources of competitive advantage, can be found in the operating core of a company, for example, at site level in contracting. *Team-specific advantages* exist in the operating core, in the bidding strategy process during the adjudication stage and also in the selection of plant and equipment and in the planning and scheduling of resources.

The preceding is arguing that the management resource at all levels is a key competitive resource. Recent empirical evidence alludes to a different picture in practice, especially when recession hits hard. During the last recession staffing cuts were both deep and wide-ranging. The result has been a change in employment patterns, with greater use of staff contracted in on a project-by-project basis and a greater acceptance of agency staff. This did not negate the need for a cadre of permanent staff to manage sites. Senior managers have always highlighted the importance of people and managers but the actions taken by them when recessionary pressures hit have implied a lack of importance of the individual except for very special members of staff. This has also been reflected in the attitudes of firms to a decline in graduate recruitment, the neglect of management development and training and the status of the personnel function. Training is a key competitive weapon in a people-intensive, knowledge-based industry and graduates provide a source of highly educated problem solvers. The human resource function appears to be one of the critical roles in continuing to develop a management cadre at all levels within the construction firm. Knowledge-based assets at all levels will be located and embedded in the distinctive capabilities stemming from organisational architecture, reputation and innovation. A disregard for investment in continually upgrading management is striking at the heart of the knowledge-based asset in firms.

The next section reviews the role of supply chain management as a source of competitive advantage.

Supply chain management in construction

The supply chain system in construction comprises two components, the demand chains of numerous clients and the supply chain of the main contractor that has to respond to a diverse range of client types, and hence demand chains. Supply chain management (SCM) is now no longer an operational issue and involves thinking about activities within the supply and purchasing function that are internal and external to the firm. Supply chains in construction can be classified as [22]:

- Professional services. These suppliers provide a combination of skills and intellectual property in the process, typically comprising the designers and other professional consultants.
- Construction and assembly – the on-site construction process – is highly skills-focused and also forms part of the service sector of the industry.
- Materials and products. These comprise the materials, products and hired plant.

Supply chain strategy for contractors is concerned with assembling the right teams, at the right times and in the right locations to manufacture a wide spectrum of products for numerous clients. Each client requires a different response from the contractor's supply chain, indicating different strategies to accommodate the needs of each. Some permit the main contractor to standardise processes and procedures and components (process and portfolio clients), others require a more or less unique response, permitting no or limited standardisation (unique and customised clients).

The main contractor acts as a supply chain hub and organises the logistics activities of Move and Store and the production activity of Make at different geographic locations. SCM could involve a closer working relationship between organisations such that for certain types of clients, the volume procurers for example, key suppliers could provide an enhanced service through greater knowledge of the client business environment and their longer term direction. This would require increased levels of trust on both sides. The adoption of formal protocols within supply chains is seen as essential. Experience from pathfinder Prime Contracts and Partnering studies also indicates that developing incentive systems through the supply chain is one of the more difficult areas.

Adopting SCM principles will also involve a cultural change and increased commitment to joint training initiatives. Supply chain strategy has also to be attuned to the different forms of procurement systems generated by the client demand chains since they have a fundamental impact on value through the management of time, cost, quality, functionality and the allocation and management of risk. Main contractors will need to develop a portfolio of supply chain strategies depending on client types, their requirements, the geographic location of projects and the product type. This may involve them in considerable process design and redesign.

Figure 10.2 sets out a framework for supply chain management within the contracting firm.

The next section explores the strategic management issues in international construction.

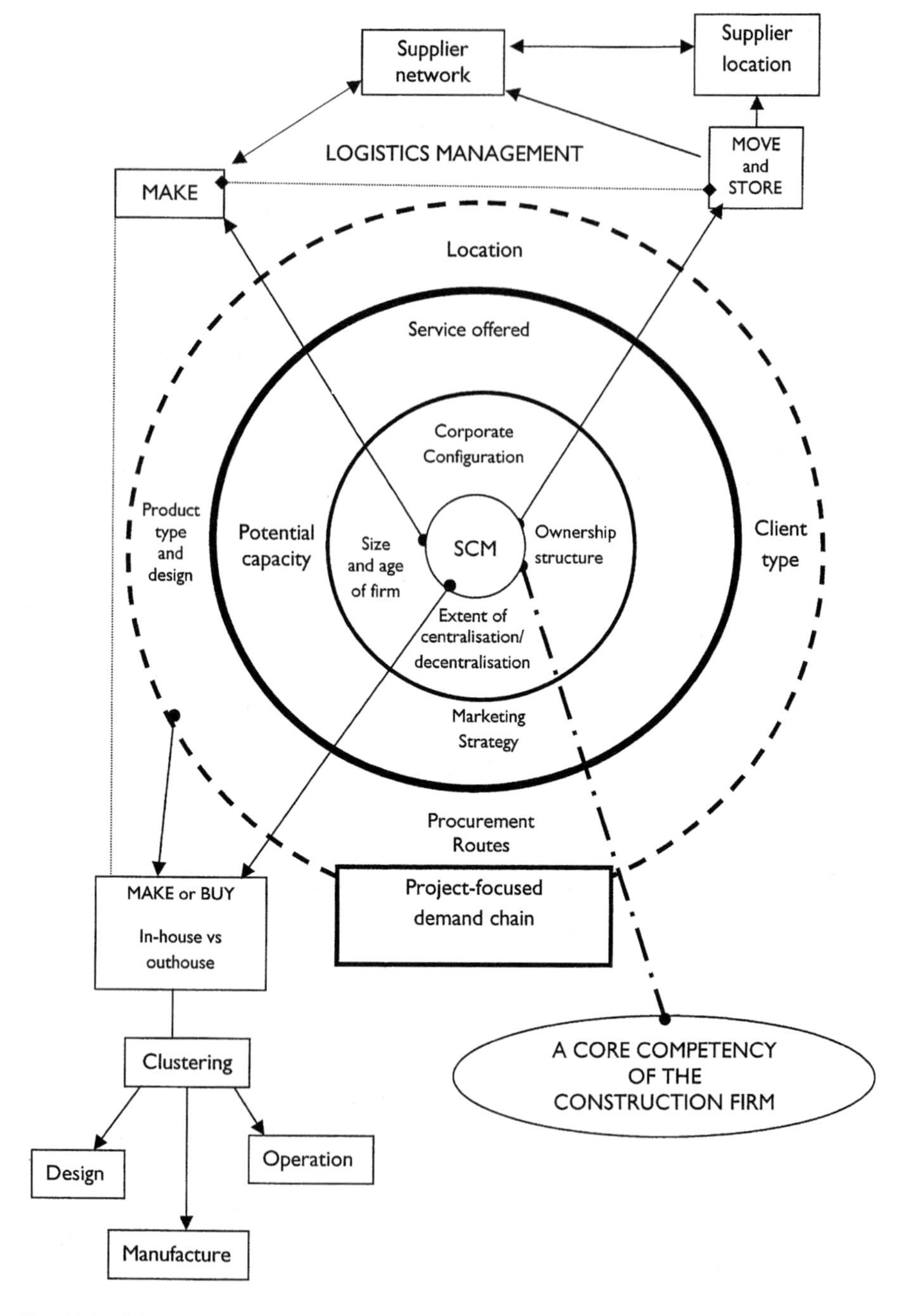

Fig. 10.2 A framework for construction supply chain management in the construction firm.

International construction

The internationalisation of UK construction firms has gained momentum during the last fifty years. UK firms are now facing corporate decisions on whether they want to become global players as a key part of their business strategy or remain within the confines of the domestic market, perhaps only venturing overseas to satisfy the requirements of key clients.

The global construction industry is large, described as fragmented and very competitive. Internationally, contractors are involved in a process of demand searching. Porter's 'national diamond' [23] and the eclectic paradigm are two frameworks that provide a powerful analytical tool for strategists. There are country-specific (CSAs) firm-specific (FSAs) and industry-specific (ISAs) advantages, forming part of ownership advantages. Country-specific advantages focus attention on host nation issues. Firm-specific advantages focus attention on those issues that are specific to a company and distinguish it from its competitors. Firm-specific advantages include system-based advantages, the fact that a firm can be located in different nations and has the opportunity to use its international organisational structure and network of activities to compete. System-based advantages in international construction included worldwide procurement of materials and information exchange. The addition of team specific advantages (TSAs) as a sub-set of firm-specific advantages also focuses attention on the importance of team-based project management skills [24]. Depending on how the project is set up and financed, project team skills become important where different specialists are drawn together:

- at project inception level
- at bidding strategy level
- at design/contract team level
- at production level in the operating core, where site management may face additionally a situation of managing a multicultural operative workforce.

In a multicultural environment, managers can [25]:

- Choose to ignore cultural differences and focus on the attainment of project and business objectives;
- Choose to take cultural background into account and design the appropriate authority structures, control systems and methods of inducement;
- Create cultural synergies by, for example, capitalising on the group orientation of a culture;
- Create a strong 'organisational culture' where the beliefs, attitudes and accepted behaviours and norms of the firm can override or modify national cultural characteristics.

Finally, there can be locational advantages (or disadvantages) of both the home and host nation.

A major review of the competitiveness of European contractors in a globalising construction market identified the international contractor's core competencies as widening and defined the emerging international contractor as one that is:

> 'a project-centred organisation able to provide flexible logistic skills, manage human resources, provide construction technical skills, organise a network of specialist, have the ability to organise and control financial packages, manage a complex multi-layered and multi-skilled organisation, which in combination can deliver an integrated global offer' [4].

The key sources of competitive advantage for international contractors are:

- The ability to provide attractive financial packages.
- The ability to build winning alliances. Alliances can be global or local, may go beyond the industry boundaries to include clients and financial institutions or they can be short-term and project-specific or longer term and whole-life focused.
- The ability to accept and manage risk.
- The ability to invest in sales and R&D.
- The identification of client/user needs through market research.
- The ability to procure on a global basis.
- Technical expertise and the right technology. The question is whether these are threshold competencies or core competencies providing a competitive advantage when unpacked against the competition. Members of the eLSEwise project were clear that technical competency would not be a source of sustainable competitive advantage amongst serious competitors for a project unless it was derived from proprietary knowledge.
- The integration of local and global knowledge.
- Political backing.

Competing internationally in construction therefore involves considering the relationships between country-specific, firm-specific, including team-specific and system-based advantages, and locational advantages.

Conclusions on managing construction corporations

The long-term subletting out of the production process combined with short-term workload variability and the successive impacts of booms and recessions has created the requirement for organisational flexibility to ensure a minimum of capital locked up in fixed and human assets. There is an unwillingness to invest in management development and training, graduate recruitment and also to sub-contract out much of the production process to others. The operating core of the contracting company is now characterised by a myriad network of sub-contractors of various types, employed with different contractual obligations, rights and liabilities, where, in addition, certain categories of sub-contractors often sublet the fixing and installation of products to self-employed specialists. Main contractors have now become resource acquirers, allocators and managers where traditional forms of contract have been at the heart of the market exchange relationship at all levels of the firm.

This chapter has highlighted the importance of developing managers at all levels to manage the firm strategically during periods of boom and recession, to manage within operating divisions and within regionally-based organisational structures and in the on-site operating core. Managers in the industry have consistently stressed the importance of people in the industry but the evidence suggests that this is not always put into practice. Construction is a knowledge-based industry, in much the same way as the information technology software industry or the engineering consulting industry. Knowledge development, maintenance and exploitation are the only sustainable source of competitive advantage for a contractor in the long term, whether locally, regionally, nationally or internationally. The empirical evidence suggests clearly that the strategic actions of managers in the industry pay scant regard to the core competencies that can be developed from knowledge assets in construction.

Hillebrandt *et al.* [3] concluded in 1995 that an elite group of large firms could emerge due to the significant capital requirements for projects. They also foresaw the possibility of the demise of the family firm. The requirement for bonding requirements on projects would also create entry barriers for many firms. Their prediction about an elite group in the industry is now well under way, with the newer procurement routes such as PFI and Prime Contracting almost certainly cementing further an already hierachical structure in the industry. The newer procurement routes will move competition to one of long-term relational contracts requiring trust and team working within supply chains. The elite group of contractors will be the core constellation of supply chain leaders and global players of the future.

The consultants – the construction professions

Not only do contractors have to think strategically but consultants have long operated in a competitive environment. The professional consultants in construction can be broken down into two broad groups: designers, namely, architects and the different specialism within the engineering profession; and surveyors, also of different specialism. For a variety of historical, status and competitive reasons the construction professions are now facing intra- and inter-professional competitive rivalry. The effects of the industry environment vary according to the size of the consultancy. Client commercial pressures in terms of greater requirements for financial, cost, legal and management control of the building process – areas where architects, for example, are under considerable threat from surveyors – are likely to exaggerate the shift towards design for architectural and engineering consultants. Clients are blurring the distinctions between professional and commercial services. Many consultancy firms will find this transition and its implications difficult to make or appreciate.

The operating environments of small practices diverge on a number of factors from those of the medium to large-sized firms and reflect the nature of the underlying power base, the nature of the client base and the strategic asset stock possessed by practices. The conditions surrounding small practices tend towards professionalism. These firms deal in small value projects and a consumer group of clients that are relatively dependent and unorganised. Moreover, professional tasks concern largely the core technical tasks and professional boundaries remain largely intact. However, as practices expand they encounter different operating environments. The larger practices operate under corporate patronage. The larger, more powerful clientele influence market conditions surrounding the high value projects. The independent and unexploitable nature of the client is presented by the willingness to procure professional services outwith the traditional professional domain, for example, from contractors. The knowledge/power base does not rest on the core professional body of knowledge but rather on sectorally specialised knowledge and management competence. No profession has managed to incorporate the 'management knowledge base' within its exclusive professional domain, nor is it likely to, given the prevalent market conditions.

The task facing construction professions concerns cannibalising valuable knowledge and techniques from other occupations while retaining a monopoly over their own [26]. The control that the professions possessed over intra-professional competition has been eroded by a number of factors. These include an over abundance of supply, the abolition of fee scales and the relaxation of marketing constraints.

Inter-professional challenges are evident over two related dimensions:

- No single profession/occupation has established supremacy in the newly emergent management services. This fact produces a complex occupational map with a variety of professional groupings tapping into this power base to extract leverage in the market place.
- The translation of the occupational knowledge base into an organisational strategy and subsequent performance in the market depends on two other streams of knowledge, namely, computer/information technology and marketing.

The professional institutions are faced with a situation where the operating conditions differ according to practice size. The need for the incorporation and development of other streams of knowledge is becoming apparent. Professional power is captured by two critical dimensions of professional service, cruciality and mystique, exercised under the prevailing socio-economic conditions found in the market place. The potency of professional power and the extent to which it can be leveraged in the market place depends on the characteristics of the clients served by a profession. As client power increases, occupational control through professionalism decreases and corporate patronage prevails. Professional power is not necessarily distributed uniformly across a profession. Different levels of professional or expert power are available to segment of professional members.

The strategic group concept provided a theoretical framework that mediates the industry and individual firm levels for analysing professional firms. In the empirical study of architects and quantity surveyors, four strategic groups were identified. The performance of the strategic groups, measured by fees generated per head of professional and technical staff varied across groups. The main influencing factors on performance were the underlying power base associated with a professional segment; the strategic assets possessed by firms and the extent to which firms pursued a competitive pricing strategy without the benefit of supporting computer technology. The implications of the empirical analysis were translated subsequently into implications for different sizes of consultancy firm.

At the level of the profession, the data clearly demonstrate a divergence between small practices and the medium/large practices in terms of the prevalent knowledge bases and the organisational strategies used to translate that knowledge into market relevance. Two major implications stem from this for the professional institutions:

- The survival and future prosperity of the professions within the different domains needs to be differentially addressed by the professional institutions.
- Practices in different environments require appropriately focused support from the respective professional associations.

Implications for professional practices and the building professions by size of firm

Small sized firms (fewer than five staff)

Small practices are operating under conditions of professionalism, although the occupational control strategies of mandatory fee scales and marketing constraints are no longer enforced. The results from this study indicate the small firm sector is the bedrock of professionalism and there is relatively little divergence from the traditional position of the professions studied, even though the occupational control strategies have been removed. Professional boundaries tend to remain intact because the services being offered are substantially based on the core professional functions. The small firm sector, as a whole, is generally under considerable financial pressure.

Small firms operate in an environment characterised predominantly by small (value) projects within the private sector. The potential breadth of strategic assets and the scope of operations are generally limited. This is as a direct result of limited resources, particularly human resources available to small firms. The majority of small firms operate within the occupational segment that depends primarily on the profession's core knowledge base for leverage in the market place (low cruciality and low mystique). A particularly significant decision for many small firms is whether it should attempt to develop resources that would permit it to operate within segments in which it may tap the power bases associated with the management service sector and/or complex innovative projects. Small firms are generally operating in an environment where fee generation potential is at a minimum and they are utilising the core knowledge base of the profession. A number of strategic choices have been identified for this size of firm:

- Developing a competitive pricing strategy supported by a productivity focused computer based infrastructure whilst seeking to develop good working relationships with industrial rather than consumer clients. The former places fewer demands on developing a comprehensive marketing strategy.
- Developing a wide ranging computer based resource and comprehensive marketing strategy aimed at highlighting the expertise of the practice in specific market sectors. This holds the best prospects for those firms that already have a favourable reputation in the marketplace.

Movement into a highly complex, innovative environment for small firms can be achieved by collaborating with firms on large projects.

Medium sized firms (6–20 staff)

The expansion from small to medium sized practices is accompanied by a move out of the professional knowledge domain and a shift from strategic groups that rely heavily on expertise and reputation to one where computer-based assets and marketing are well developed. While the small firms are concentrated within the segment where the knowledge base is associated with the core professional body of knowledge, the medium sized firms are distributed over a wider range of segments. This indicates that as firms expand in size, they develop services/assets that permit them to tap the power bases associated with management services and complex/innovative projects. They also seek higher project value environments as they move out of the 'low mystique and low cruciality' segment associated with small value projects. A particular problem faced by the medium sized firms operating this environment is that because projects tend to be small, a higher number of commissions have to be secured to generate the same fees per head ratio.

For medium sized firms the best prospects for development involve moving out of the traditional professional knowledge base and into the provision of management services. This will, however, move them into a highly competitive environment and they will need, at the same time, to develop competitive pricing strategies, supported by a computer-based asset, and a comprehensive marketing strategy. Medium sized firms operating on highly complex/innovative projects do not tend to achieve superior performance results.

Large firms (more than 20 staff)

The large firms congregate around segments with power bases established on management services and tend to be the highest performers in terms of fees generated. The importance of having a wide array of strategic resources in terms of computer-based assets, sectoral specialisation and marketing is reflected in the distribution of large firms among the strategic groups.

As large practices venture into environments where projects are complex or require high levels of creative input the performance results tend to drop. This is attributed to the higher levels of senior professionals' effort and time spent in solving complex problems. It is in this area that the presence of competition from management contractors and construction managers is being felt the most. This is supported by comments made by Latham [27] is his wide ranging review of procurement and contractual arrangements in the UK construction industry. The rise of contractor-led procurement has opened a window of opportunity for contractors to leverage their knowledge/power base associated with management of the project development process, particularly in the complex/innovative projects requiring crucial management input.

Conclusions on managing construction consultancies

Medium and large practices tend to operate in environments characterised by corporate patronage, including private clients commissioning large projects, the public sector and contractors. These are the experienced clients and the data indicate they are neither exploitable nor dependent on professionals. These clients have choice that can be exercised in the market place for the emergent services. The extensive use of marketing indicates that large practices have made full use of the relaxation of occupational controls. 'Client care' is essential for these firms. They are operating in a very competitive arena where a range of strategic assets are required for survival and continued development.

The data from this study indicates that the large quantity surveying and architectural practices compete directly in the management sector, where they will also encounter competition from the contractors. Within this environment, not only are occupational control strategies dissolved but also professional boundaries are no longer distinct. The development of a body of knowledge concerning management-based services is increasingly becoming vital to the continuing professionalisation process.

The data presented here are also clear. There are different bodies of knowledge within the professions that permit leverage in the market place in different ways through the translation of this into competencies. Management-based knowledge and sectoral specialisation are considered bodies of knowledge that define the emerging power bases and are rooted in the construction and property fields. A number of issues stem from this for the professional institutions:

- The manner in which these emergent bodies of knowledge should be developed and incorporated within the professions' bodies of knowledge and the way they should be imparted in formal training.
- The development of a framework within which they may be developed in current practising firms, especially the smaller firms that wish to expand.
- The manner in which competence and specialisation in these fields may be recognised within the professional credential system.
- The marketing of the professions' new competence in the new areas.

Marketing management has been identified as significant in translating the underlying professional knowledge base into market relevance. Computer-based knowledge has a dual role: first, its development within the occupational body of knowledge as a catalyst for new services using the technology; second, its application at the organisational level for increasing productivity and to support a competitive pricing strategy.

Finally, the professional institutions will need to consider their role as the repository, tester and qualifier of knowledge for practice. This should involve considerations of the extent to which they embrace the bodies of knowledge identified above to support membership in the short term in the context of organisational strategies and in the longer term for occupational strategy. This is in addition to the development of the core body of professional knowledge.

The next section reaches a series of conclusions about strategic management in construction and develops a managerial framework to assist managers in the strategic management process.

Conclusions

This chapter has drawn together a number of major themes running through the book. Its primary focus has been to place construction firms in the context of theories of organisation, organisational behaviour, economics and corporate strategy, with a view to developing insights into how to achieve competitive advantage in construction.

One thing that has become abundantly dear is that the terminology used to describe construction firms needs to be more precise. The larger construction firms have now become so diversified that contracting forms only a part of their overall operations. Hillebrandt & Cannon [28] suggested using 'construction firm' to refer to the diversified company operating in the construction industry, and 'contracting firm' to refer to a company operating purely to erect or modify structures. This has been elaborated further in this book. Where possible, reference has been made to those issues that will affect an analysis of the diversified construction company. It has also become abundantly clear that theories of 'construction companies' now need to be developed. The analysis presented in this and earlier chapters indicates that construction firms need to be close to the market, both in terms of client and consultant contacts, and for more detailed knowledge of labour conditions and localised economic environments. The strategic management process in construction companies should be a combination of both 'bottom-up' and 'top-down' procedures: 'bottom-up' so that strategists can incorporate localised market knowledge into their thinking in order to remain in contact with the market place, in general, and ensure that the strategic gestalt is maintained; 'top-down' to ensure that the vision and direction of the company is communicated throughout the firm to enable a common direction and purpose to be developed or maintained. This approach to strategic management should facilitate the 'loose–tight' style of management. Major international forces are now at work in the environment that will have a significant impact on domestic construction industries and those firms operating internationally.

Equally, since deregulation, consultancy firms have also become major players in offering a wide range of services, many of which bring them into direct competition with each other or construction firms. They are also facing significant pressures for change on their traditional roles in the industry.

A contingency model of strategic management for construction

Figure 10.3 sets out the basic framework of the contingency model of strategic management in construction, drawing together the important issues from each chapter.

The model can be used as the first step in the strategic analysis of a firm, making reference to the relevant sections in each chapter for understanding the components of the analysis. The depth to which the analysis is taken depends very much on the size of firm and the resources available for the analysis. In conclusion, Clark [29], using empirical research, asserts that large firms in particular have repertoires of multiple structures. The analysis of the evolution of the construction industry since the 1950s has suggested that different environmental conditions have emerged, requiring different responses from senior managers and their organisations. Firms, having faced such different environmental conditions, would have stored within their collective experience different structural repertoires. Some organisational structures would lie dormant for long periods of time to re-emerge when required by forces within the environment and/or within the firm. Lansley [30] has argued that as a consequence of a downturn in economic activity there are environmental pressures on firms to rationalise and hence adopt innovation and new value systems ready for the next upswing in economic activity.

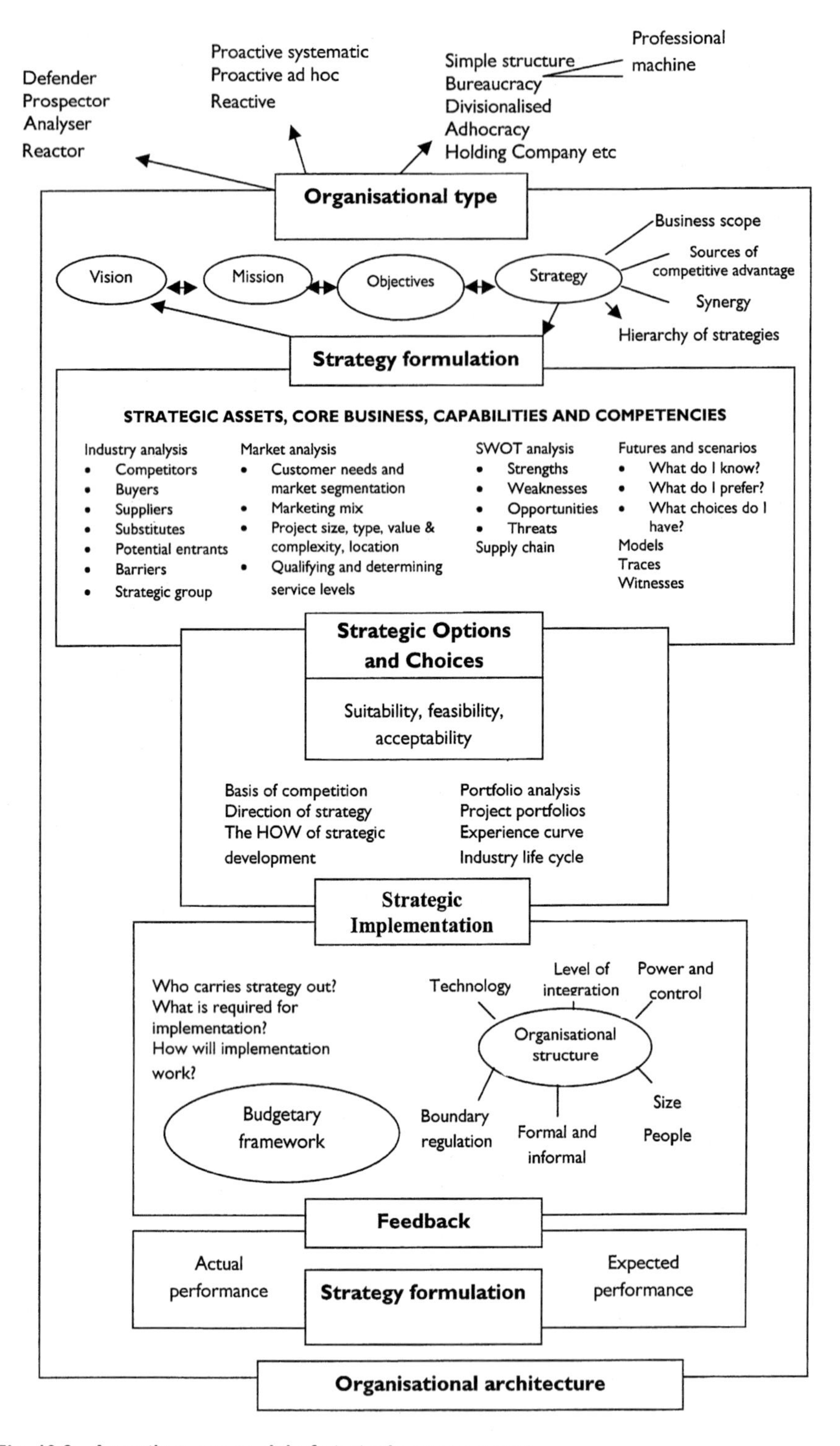

Fig. 10.3 A contingency model of strategic management.

A number of typologies have been put forward for describing firms in the construction industry. Reference has been made to some of these throughout the preceding chapters. However, the typology suggested by Mintzberg [7] is useful in moving towards the implications for strategic management in construction and this is outlined below.

Simple structure

This structure is endemic in the construction industry. It is typified by minimal structure and is organic. The problem for this organisational type is that the operating and strategic decision-making processes often become intermixed and overlap. As a result of organisational size, the strategic management process if present in any identifiable form, will probably be incremental, short term and reactive. The simple structure characterises many sub-contracting firms.

Simple professional structure

This structure is a characteristic of the small consultancy organisation. Small architectural practices have been identified as being highly organic. This stems from the fact that they face environments that are dynamic because of the volatility of workloads.

Professional bureaucracies

As consultancy firms become larger the emergence of this structure occurs where the emphasis for the production level is on standardised skills. This structure typifies the medium and large consultancy firms prior to the removal of scales of fees. In this respect, the environments faced by these firms were relatively predictable since they were likely to have a more stable clientele and revenue generation was controlled. However, due to the introduction of competitive fee bids and the fact that client pressures are changing, the professional bureaucracy may give way to a number of different structural forms in the professions. In this instance the pull by market forces could be towards adhocracy or machine bureaucracy depending on the degree of personal services or technology emphasised in the production core and the degree to which the control over standardisation is internally or externally generated. However, the important point is that for consultancy firms moving away from the professional bureaucratic form the strategic management process would be different in each case. For the consultancy firm investing heavily in technology to gain a competitive edge, the strategic management process would favour the systematic planning mode to enable the firm to keep abreast of technological developments. For the consultancy firm moving towards adhocracy, where the emphasis is on a personalised client-oriented service, the emphasis would favour the systematic ad hoc mode.

Depending on their size, construction corporations will have the most complex structural forms in terms of Mintzberg's typology. The large, multiple SBU construction organisations will have a *divisional structure*. However, the nature of each division will be different. For example, in those construction companies with divisions that cover building materials (machine bureaucracy), property development and project management (professional bureaucracy/ adhocracy) and contracting (adhocracy), the strategic management process becomes one of managing divisional interdependencies and disparities between market structures.

The tracing of the environmental evolution of the construction industry since the Second World War has highlighted the fact that familiarisation of present and future environments is assisted by a formal strategic planning process and the development of a clear vision of the future direction of the firm, its objectives and a set of strategies to meet those objectives. Chapter 8, which looked at futures, highlighted the fact that underpinning this work is uncertainty and a philosophy of change. Ansoff [9] has argued that under a given set of environmental conditions ranges of strategic behaviours are possible. The message conveyed here is that regardless of the size of a firm, strategic decisions can be made that increase the degree of 'fit' between the firm and its business environment and hence improve its chance of survival. The examination of futures highlighted three fundamental questions that can be addressed regardless of firm size. To reiterate, these were:

- What choices do I have? This question focuses the mind on constraints and associated resource implications. Again, this has to take account of organisational politics.
- What do I know? This question firmly addresses the current knowledge base of the firm. A range of techniques can be used to address this issue of a firm's capabilities.
- What do I prefer? Preferences usually involve some political consequences within organisations.

The size of firm will obviously have an impact on the degree to which futures work can subsequently be used. For example, the use of models, traces or analysis of witnesses to create alternative possible futures would be beyond the scope of many organisations within construction. However, this does not negate the principles upon which it is based. For example, the board of directors of a medium-sized contracting or consultancy firm could generate a series of possible futures based on their own knowledge of the industry. The choice of perhaps a couple of probable futures would sensitise senior management to a range of issues that the firm may not have considered. If a series of alternative strategies are developed for these probable futures, the firm has moved towards second generation or contingency planning. The problem associated with exploring and examining possibilities, which the implementation of futures work entails, is that it involves threats and challenges to existing ways of thinking and modes of behaviour. There is no reason why a 'futures' type of analysis could not be undertaken in either the systematic planning or systematic ad hoc modes.

The generation of alternative courses of action, or as the review of futures work has suggested, of alternative possible futures, will facilitate flexibility. The key issue is that a firm's senior managers, as the strategic decision-makers, have a commercial role to play in shaping the flexible firm. Of the five criteria isolated and suggested by Lansley *et al.* [31]. for strategic flexibility, three are directly related to the actions of senior management and the other two stem from such action, with their effects being felt by employees.

The crux of this analysis is that the construction industry is heavily dependent on human capital. In practice this means, for a firm:

(1) using the knowledge the firm owns. This knowledge may be held by individual employees or embedded in procedures or documents
(2) the firm's ability to learn from the past in order to be better able to predict the future
(3) investing in training or management development
(4) evaluating the firm's vulnerability to the erosion of knowledge through staff turnover.

To achieve these factors and translate them into strategic management behaviour it is likely that the firm will need to co-operate with other in the market to achieve competitive advantage. The watchword is competition through co-operation.

References

[1] Porter, M.F. (1980) *Competitive Strategy: Techniques for Analysing Industries and Competitors*. The Free Press, New York.
[2] Male, S.P. (1991) Strategic Management in Construction: Conceptual Foundations. In: *Competitive Advantage in Construction*. S.P. Male & R.K. Stocks (eds), pp. 5–44. Butterworth-Heinemann, Oxford.
[3] Hillebrandt, P.M., Cannon, J., Lansley, P. (1995) *The Construction Firm In and Out of Recession*. Macmillan, Basingstoke.
[4] Male, S.P. & Mitrovic D. (1999) Trends in World Markets and LSE Industry. *Engineering Construction and Architectural Management*, **6**, No. 1, 7–20.
[5] Kay, J. (1993) *Foundations of Corporate Success: How Business Strategies add Value*. Oxford University Press, Oxford.
[6] Lansley, P.R. (1991) Organisational Innovation and development. In: *Competitive Advantage in Construction*, S.P. Male & R.K. Stocks (eds), pp. 128–138. Butterworth-Heinemann, Oxford.
[7] Mintzberg, H. (1979) *The Structuring of Organisations*. Prentice Hall Inc., NJ.
[8] Miles, R.E. & Snow, C.C. (1978) *Organisational Strategy, Structure and Process*. McGraw-Hill, Tokyo.
[9] Ansoff, I. (1987) *Corporate Strategy*, 2nd Edn. Penguin, Harmondsworth.
[10] Vasconcellos, E. (1999) *The War Lords: Measuring Strategy and Tactics for Competitive Advantage in Business*. Kogan Page, London.

[11] Mintzberg, H., Quinn, J.B., Ghoshal, S. (1999) *The Strategy Process*. Prentice Hall, Harlow.
[12] Mintzberg, H. (1987) Crafting Strategy. *Harvard Business Review*, July–August.
[13] Stocks, R.K. (1991) Strategic marketing management. In: *Competitive Advantage in Construction*, S.P. Male & R.K. Stocks (eds), pp. 105–127. Butterworth-Heinemann, Oxford.
[14] Thorpe, T. & McCaffer, R. (1991) Competitive bidding and tendering policies. In: *Competitive Advantage in Construction*, S.P. Male & R.K. Stocks (eds), pp. 163–194. Butterworth-Heinemann, Oxford.
[15] Ohmae, K. (2000) *The Invisible Continent: Four Strategic Inspirations of the New Economy*. Nicholas Brearly, London.
[16] Ibrahim, A.A.R. (1995) *An Empirical Analysis of the Strategic Group Concept Within the UK Construction Industry*. Unpublished PhD, Heriot-Watt University, Edinburgh.
[17] Male, S.P. (1991) Strategic management and competitive advantage in construction. *Competitive Advantage in Construction*, S.P. Male & R.K. Stocks (eds), pp. 45–104. Butterworth-Heinemann, Oxford.
[18] Wrigley, L. (1970) *Divisional Autonomy and Diversification*. Unpublished DBA Thesis, Harvard Business School, Cambridge, MA.
[19] Rumelt, R.P. (1982) Diversification strategy and profitability, *Strategic Management Journal*, **3**: 359–369.
[20] Waterfall, M.H.P. (1989) *The Profitability of Labour Only Sub-contracting*. MSc Thesis, Heriot-Watt University, Edinburgh.
[21] Cheetham, D.W. (1982) Labour management practices. *Construction Papers I*. No. 3, 37–53.
[22] *Croner's Management of Construction Projects*, July 1999, pp. 2-542–2-543.
[23] Porter, M.E. (1990) *The Competitive Advantage of Nations*. Macmillan, London.
[24] Enderwick, P. (1989) Multinational contracting. In: *The Multinational Service Firm*, P. Enderwick (ed.). Routledge.
[25] Tayeb, M. (1991) The effects of culture on the management of organisations. In: *Competitive Advantage in Construction*, S.P. Male & R.K. Stocks (eds), pp. 236–247. Butterworth-Heinemann, Oxford.
[26] Spiteri, J. (1999) *A Critical Analysis of Occupational and Organisational Strategy in UK Architectural and Quantity Surveying Practice*. Unpublished PhD. University of Leeds.
[27] Latham, M. (1994) *Constructing the Team*. Final Report of the Government/Industry Review of Procurement and Contractual Arrangements in the UK Construction Industry. HMSO, London.
[28] Hillebrandt, P.M. & Cannon, J. (1990) *The Modern Construction Firm*. Macmillan, Basingstoke.
[29] Clark, P. (1989) Social technology and structure. In: *The Management of Construction Firms: Aspects of Theory*. P.M. Hillebrandt & J. Cannon (eds), pp. 75–91. Macmillan, Basingstoke.
[30] Lansley, P. (1987) Corporate Strategy and Survival in the UK Construction Industry, *Construction Management and Economics*, **5**: 141–155.
[31] Lansley, P., Quince, T., Lea, S. (1979) *Flexibility and Efficiency in Construction Management*. Final Report. Building Industry Group, Ashridge Management College, Amersham, Bucks.

Index

acquisitions
  and business environment 29–30
  management resource 118
  portfolio management 161
  for strategic development 78
adhocracies 85, 219, 239
alliances *see* partnerships
analyser structures 86
Ansoff typology, firms 87, 204
architects
  business environment 35, 36–7, 38, 39, 231–2, 235
  organisational structures 83
  portfolio management 160–61

Boston Consulting Group (BCG) growth–share matrix 157–9
branding 6
business environment 1–3, 196–7
  actors in 28–39
  generic business idea 168
  international 132–5, 144, 147–9
  market overview 1–2, 11–27
  market research 181
  professionals 35–9, 231–3
  recurrent change 196
  scenario planning 167–8
  strategic decision-making contexts 42–54
  theory–practice synthesis 6, 195–7
    construction firm management 216–31
    contingency model 237, 239, 240
    professions 231–3
  transformational change 196
business portfolio management 157–61, 215–16
business unit strategy 68, 100, 107, 108–11, 205, 222
  contingency model 240
buyers, and industry structure 44
  see also clients

capital requirements
  construction strategies 220
  elite firms 231
  as entry barriers 49–50, 199
  operating core 111–12
  and strategy development 103, 104
centralisation 64, 107–108, 110–11, 200, 222
change management 80–83, 109, 209–10
civil engineering 34, 39
clients 42
  contractual arrangements 47–8, 54
  distribution channels 50, 54, 197–8, 199
  and industry structure 44
  industry/market distinction 52–3, 197
  international construction 148
  marketing 176–80, 181–2, 186–8, 211–14
  procurement options 46, 50, 53–4, 197–8
  product differentiation 48–9
  product life-cycle 43
  project life-cycle 43, 53–4
  promotional strategies 184, 185, 188
  service quality 177–80, 186–8, 212, 213
  switching costs 50, 199
  tendering strategies 47, 54
commercial building 13–14, 20–21, 23, 24
common industry environment 30–31
competencies, core 97–9, 120, 136, 138, 202, 230
competition
  industry structure 11–12, 43–5
  international market 134, 135–44, 148–9, 150–51
  professions 36, 37, 38–9, 232–3
  strategic options 77–8
  tendering strategies 47, 49
competitive advantage 69–75
  competitive strategy 69, 210–11
  core business 201
  core competencies 97, 120, 136, 138, 201, 230
  innovation 71–5, 201, 202–3
  international market 135–44, 147, 229, 230
  management resource 115–19, 225–6
  scenario planning 167–70
  site management 119–20, 225

competitive change 196, 203, 209
competitive environment 31
competitive intelligence 150–51
competitive strategy 69, 210–11
competitor analysis 76, 214
complexity, organisational 63, 200
construction corporations, management 216–31, 237
construction firms, terminology 236–7
construction industry, classification 32–3
construction strategies 220–23
consultants
  as business organisations 3–4, 6–7, 38–9, 196–7, 231–6, 239
  change management 83
  as distribution channels 54, 197–8, 199
  innovation 74
  international 127–8, 148
  portfolio management 160–61
  service quality 187
  strategic behaviour 93
  structure of professions 35–9
contestable markets 26, 200
contingency model 237–41
contracting corporations 217, 219
  see also contractors
contractors 32–5, 45–6
  buyers' procurement strategies 46, 50, 53–4, 197–8
  buyers' switching costs 50, 199
  capital requirements 49–50
  competitive advantage 115–17, 120, 211
  construction strategies 220–23
  contractual arrangements 47–8, 54
  distribution channels 50, 54, 197–8, 199
  elite group 120–21, 231
  entry and exit barriers 48–52
  experience curve 51, 199
  industry/market distinction 52–3
  innovation 72, 73, 74
  international *see* internationalisation
  management resource 115–17, 120, 225–6
  portfolio management 160
  post-war evolution 28–31
  pre-qualification 47, 48–9, 187
  product differentiation 48–9, 52, 220–21
  project portfolios 112–15, 223–4
  scale economies 51
  service quality 187
  strategic choice 105
  supply chain management 116–17, 226–8
  tendering strategies *see* tendering strategies
contracts 47–8, 54
  psychological 219
  relational 94, 96, 117, 201, 213, 225
  sub-contractors 114, 224
  value chain analysis 70–71, 215
core business 94–7, 98–9, 201–202
core competencies 97–9, 120, 136, 138, 202, 230
core products 97–8
corporate strategy 68, 99, 205
cost leadership 77, 78
country analysis 144–6
cross impact analysis 171–3
cultural factors
  internal 178–9, 187, 197, 213–14
  internationalisation 134–5, 229

decentralisation 64, 107–108, 110–11, 200, 222
decision making
  levels of strategy 100
  models of the firm 61
  regional structures 107, 108, 222
  and strategic change 81
  strategists' role 84, 208–10
decisions, types of 61–2
defender structures 86
Delphi techniques 5, 162–6
designers
  innovation 71–2, 74
  international 127–8
determining levels of service 47, 49
differentiation strategy 77, 78, 220–21
distribution channels 50, 54, 197–8, 199
diversification 78, 79, 207
  available strategies 221–2
  business environment 29–30, 32, 34–5, 38–9
  construction corporations 216–18
  entry and exit barriers 49–50, 51, 52, 199
  financial management 102–103
  international *see* internationalisation
  management resource 118
  reasons for 106–107, 221
  strategic management 100–102, 219–20, 236–7
    centralisation 107–108, 110–11, 222
    change 104–106, 109
    financial drivers 102–104, 107, 111–12
    management resource 115–19, 120, 225–6
    operating core 111–15, 219–20, 223
    project portfolios 112–15, 223–4
    regionalisation 107, 108–11, 222

sub-contracting 113–15, 224–5
supply chain 116–17, 226–8
divisional structures 86, 222
levels of strategy 99–100
portfolio management 159–60, 215–16
domestic demand conditions 138, 142

eclectic paradigm 141, 143
economic factors, internationalisation 134, 140–41, 144–5
economies of scale 51, 199
elite firms 120–21, 231
eLSEwise project 136, 137, 138
emergent strategy 197
employee relations
marketing 178–9
operatives 219–20, 225
end-products 97–8
engineers
business environment 34, 38, 39, 232
portfolio management 160–61
entry barriers 44–5, 48–52, 198–200
exit barriers 45, 48–52, 198–200
experience curve 51, 199

factor conditions, internationalisation 138, 141, 143
financial factors
construction strategies 220, 221
elite firms 231
as entry barriers 49–50, 199
internationalisation 134, 137, 147–8
operating core 111–12
strategy development 102–104, 107
firm, the 59–65, 200
see also organisational structure
fiscal factors
employment status 24, 113
international construction 148
focus (niche) strategy 77–8
foreign direct investment (FDI) 140–41, 143
formal structure, firms 64
formalisation, organisational structure 64, 200
functional (operating) strategy 68, 205

general environment 28, 31
generic business idea 168
geographical factors
management resource 118
market analysis 17–18, 26, 198
regional structures 108, 109, 222
see also internationalisation
global construction firms 136
global market analysis 125–9, 130, 135
government policy
as entry barrier 51, 199
internationalisation 140
and organisational structure 93–4
group-think 84
growth strategies, regionalisation 108, 222

holding company structure 88
horizontal differentiation 63, 200
housing 14–16, 19, 22, 24
human resource function
management resource 117–18, 119, 226
and market trends 24
training and development 49, 117–18, 119, 120, 213, 226
wage structure 220
human resources 24, 115–19, 120, 224–6, 241
marketing 182–3, 213

industrial building 12–13, 20–21, 22–3, 24
industry, market compared 45, 52–3, 197
industry life cycle 42–3
industry structure
competitive strategies 43–5
contractors 32–5
demand variability effects 11, 26, 198, 219
elite group 120–21, 231
entry barriers 44–5, 48–52, 198–200
exit barriers 45, 48–52, 198–200
forces shaping 44–5, 52–3, 197–8
fragmentation 2, 11–12, 26, 93, 199–200
and organisation structure 92–3
post-war evolution 28–32
informal structure, firms 64, 200
innovation 71–5, 201, 202–203
core business 97, 99, 201
institutional level, firms 59–60
internationalisation 4–5, 123–4, 229–30
business environment 132–5, 144, 147–9
competitive advantage 135–44, 147
core competencies 136, 138, 230
domestic demand conditions 138, 142
entry barriers 51
factor conditions 138, 141, 143
firm strategy 138, 142–4
firm structure 138, 142–4
management resource 118, 119
market analysis 125–9, 130, 135

market diversification 34–5
obstacles 133–5
reasons for 129–32
scenario planning 170–71
for strategic development 79
strategy 144
competitive intelligence 150–51
country analysis 144–6
general overview 146–9
sub-contractors 138, 142
suppliers 138, 142, 143

joint ventures 161

knowledge integration 138

labour *see* operatives
language, international construction 147
legal factors
international construction 148
scenario planning 169
liquidity, financial 102–103
local authorities, housing 15, 17

machine bureaucracies 85, 239
management control, internal 118–19
management development 117–18, 119, 120, 226
management hierarchy
models of the firm 59–61, 62, 64
organisational change 83, 104–105
and psychological contract 219
regional structures 110–11
management resource 115–19, 120, 225–6
management services 35, 234–5
manufacturing industry, construction as 33, 34
market(s)
BCG growth-share matrix 157–9
industry compared 45, 52–3, 197
international 4–5, 125–9, 130, 135
see also market analysis; market structure
market analysis 1–2, 11–12
commercial building 13–14, 20–21, 23, 24
contestable markets 26, 200
demand variability effects 11, 24–6, 198
environment-structure link 6, 197–8
financial factors, capital 24
geographical factors 17–18, 26
housing 14–16, 19, 22, 24
indicators 17
industrial building 12–13, 20–21, 22–3, 24
and industry structure 2, 11–12, 26, 197–8, 199–200
market areas 11
public sector 15, 17, 18–20, 22, 23
repair and maintenance 2, 17, 18–20
market research
international 137
for marketing 181, 187, 213
market structure 28–32
capital requirements 49–50
contractual arrangements 47–8, 54
distribution channels 50, 54, 197–8
entry and exit barriers 48–52, 198–200
experience curve 51
government policy 51
and industry structure 44–5, 52–3, 197–8
and organisation structure 92–3
procurement strategies 46, 50, 53–4, 197–8
product differentiation 48–9
professions 35–9, 232–6
project-based 26
scale economies 51
switching costs 50
tendering strategies 47, 49, 54
types of 45–6, 197–8
marketing 6, 175–6, 211
customer care 179–80, 182–3, 186–8, 213
customer orientation 176–7, 211–12
customer satisfaction 177–8, 179, 212
definitions 176
front-line staff 182–3, 213
function 175
interactive 179
internal 178–9, 183
key client 181
mix 183, 212–13
philosophy 175, 211–12
by professions 235, 236
research 181, 213
segmentation 180–81, 212
service quality 177–80, 186–8, 212, 213
strategies 175, 180–83, 213–14
materials
costs of 111
innovation 72, 73, 203
matrix analyses, portfolio management 157–9
matrix structures 88
media, promotional 184–5
mergers
and business environment 29–30
management resource 118

portfolio management 161
for strategic development 78
Miles-Snow typology, firms 86, 204
Mintzberg typology, firms 84–6, 203–204, 219, 239–40
mission statements 68
multinational companies (MNCs) 88
multinational construction firms 136
multinational enterprises (MNEs)
foreign direct investment 140–41, 143
structure 88

national environment 30–31
niche markets, international 128–9
niche strategy 77–8

objectives of firms 68–9
internationalisation 131
on-site production *see* operating core
operating core 119–20
competitive advantage 116–17, 119–20, 226
major inputs 111–12, 223
management resource 116–17, 120, 226
model of the firm 60–61
operatives' attitudes 219–20, 225
potential capacity 113, 219, 224
project portfolios 112–15, 224
strategic decision making 100
sub-contracting 113–15
operational environment 31
operational strategy 68, 205
operatives
attitudes of 219–20, 225
employment status 24, 113, 219, 220
production process 111, 112
skill levels 32, 224–5
sub-contracted 113, 224–5
wage structures 220
opportunities, SWOT analysis 75, 214
organisation, the, models 59–65
organisational architectures 94, 96, 97, 99, 120, 201
organisational change, management 80–83, 109, 209–10
organisational innovation 72–4, 203
organisational level, firms 60
organisational structure 63–4, 200
Ansoff typology 87, 204
change management 82–3, 109
construction industry 92–4, 219–20
centralisation 107–108, 110–11, 222
core business 94–7, 98–9, 201–202
core competencies 97–9, 202
regionalisation 108–11, 222
strategic choice 104–105
holding company 88
internationalisation 138, 142–4
levels of strategy 99–100
matrix 88
Miles-Snow typology 86, 204
Mintzberg typology 84–6, 203–204, 219, 239–40
multinational 88
sectoral 87
and strategic implementation 79
subsidiary company 88
virtual enterprise 88–9

partnerships
international construction 137, 148
portfolio management 161
people, marketing mix 183, 213
see also human resources
personnel function *see* human resource function
physical system, firms 64–5
place, marketing mix 183, 213
planning process, strategic management 101–102, 206
see also strategic planning techniques
politics
housing market 15–16
internationalisation 133, 134, 138, 144, 145–6
Porter's national diamond 138–40, 141, 144
portfolio analysis 129–30
portfolio management 157–61, 215–16
potential capacity 113, 219, 224
power culture 197
preliminaries estimate 70, 71, 215
price
experience curve 51
marketing mix 212–13
product differentiation 49
tendering strategies 47, 49, 117, 186, 212–13, 224
value chain analysis 70, 71
Private Finance Initiative (PFI) 42, 221
proactive adhoc behaviour 87
proactive systematic behaviour 87
process innovation 73, 74, 203
procurement 46, 50, 53–4, 197–8
elite contractors 121
innovation 74

international construction 137, 146
product, marketing mix 183, 212
product differentiation 48–9, 52, 199, 220–21
marketing 187, 188
product innovation 73–4, 203
product life cycle 42–3, 201
production, on-site *see* operating core
professional bureaucracies 85, 239
professions/professionals
business environment 35–9, 231–3
as business organisations 3–4, 6–7, 38–9, 231–6, 239
firm size 233–5
organisational structures 83, 239
portfolio management 160–61
project-based market structures 26
project life cycle 43, 53–4
construction strategies 221
operatives' attitudes 219–20
project management services 35, 234–5
project portfolios 112–15, 160–61, 223–4
project teams 116, 226, 229
promotional strategies 6, 176
marketing mix 183, 213
media 184–6
objectives 183–4
tender shortlisting 185–6
prospector structures 86
psychological contracts 219
public sector construction 15, 17, 18–20, 22, 23

qualifying levels of service 47, 49, 202
quality
customer service 177–80, 186–8, 212, 213
generic business idea 168
quantity surveyors 35, 36, 37, 235

reactive behaviour 87
reactor structures 86
regional structures 100, 107, 108–11, 222
relational contracts 94, 96, 117, 201, 213, 225
relationship marketing 177
repair and maintenance (R&M) 2, 17, 18–20, 32
reputation
core business 96–7, 201
product differentiation 48, 49
promotional messages 185–6
scenario planning 168–9
risk management, international 137
role perspective, firms 64–5, 200

scale economies 51, 199
scenario planning 5–6, 166–73
sector structure 87
selling-oriented organisations 177, 212
service
determining levels of 47, 49
marketing strategies 177–80, 186–8, 212, 213
qualifying levels of 47, 49, 202
service industry
construction as 33–4
marketing mix 183
services engineers 39
site management 111, 119–20, 219–20, 224, 225
situational perspective, firms 64–5, 200
social processes
firms 64, 200
housing market 15–16
organisational architecture 96, 201
spatial dispersion, organisations 63, 200
specialisation 28–9, 30, 198, 221
professions 35, 36–7, 38
speculative construction 3, 46, 198
strategic behaviour 83–9, 92–4
contingency model 240–41
core business 94–7, 98–9
core competencies 97–9, 120
levels of strategy 99–100
see also strategic management process, construction firms
strategic business areas (SBAs) 77, 215
strategic business units (SBUs) 68, 222
contingency model 240
levels of strategy 100
portfolio management 157–9, 215–16
regionalisation 108, 222
strategic change 81, 196, 203
strategic choice 67, 77–9, 207, 215
construction firms 104–106
financial drivers 102–104, 107
management resource 115–19
professional practices 234
strategic development
financial drivers 102–104
means of 78–9
strategic feedback 79, 80
strategic formulation 67–76, 206–207, 214–16
strategic groups 45, 76, 199, 214–15, 233
strategic implementation 67, 79–80, 207–208
strategic level, firms 59–60
strategic management process 3–4, 67, 206
change 80–83, 104–106, 109, 209–10

construction firms 100–102, 219–20, 236–7
centralisation 107–8, 110–11, 222
change 104–106, 109
contingency model 237–41
financial drivers 102–104, 107, 111–12, 221
international *see* internationalisation
management resource 115–19, 120, 225–6
operating core 111–15, 219–20, 223
project portfolios 112–15, 223–4
reasons for diversification 106–107
regionalisation 107, 108–11, 222
sub-contracting 113–15, 224–5
supply chain 116–17, 226–8
strategic choice 67, 77–9, 102–106, 107, 115–19, 207, 215
strategic feedback 79, 80
strategic formulation 67–76, 206–207, 214–16
strategic implementation 67, 79–80, 207–208
strategists' decision-making 84, 208–10
strategic planning techniques 5–6
Delphi techniques 5, 162–6
portfolio management 157–61, 215–16
scenarios 5–6, 166–73
strategic positioning 150
strategic square 215
strategists' role 84, 208–10
strategy
definitions 65–7, 204–206
internationalisation 144–51
levels of 99–100
seven theories of 116
types of 68
strengths, SWOT analysis 75, 214
structural engineers 39
sub-contracting 113–15, 219
business environment 30, 38
and innovation 72, 73, 203
international construction 138, 142
management resource 120, 225
value activities 70
subsidiary structures 88
levels of strategy 100
portfolio management 159–60, 215–16
substitute products 45, 53–4
substitute services 45, 53–4
suppliers
industry structure 44
internationalisation 138, 142, 143
supply chain management (SCM) 116–17, 226–8
elite groups 121
on-site 225
surveyors
business environment 35, 36, 37, 38, 39, 231–2, 235
portfolio management 160–61
switching costs 50, 199
SWOT analysis 75, 214

takeovers
business environment 29–30
management resource 118
portfolio management 161
for strategic development 78
task environment 28, 31
team-specific competitive advantage 116, 226, 229
technical expertise, international 137–8
technical system
firms 64–5, 200
innovation 72, 74
technology
innovation 72, 73–4, 203
international construction 137–8
scenario planning 169–70
technical system compared 65
Technology Foresight programme 165–6
tendering strategies 47, 49, 54
competitive advantage 70–71, 116, 117, 215
contestable markets 26, 200
emergent 197
internationalisation 146
management resource 116, 117
marketing 187, 212–13
price 47, 49, 117, 186, 212–13
promotional messages 185–6
sub-contractors 224
value chain analysis 70–71, 215
threats
of entry 44–5
SWOT analysis 75, 214
time horizons
Delphi technique 164–5
strategies 99, 206
trade contracting *see* sub-contracting
trade restrictions 133–4
trade unions 225
training 49, 117–18, 119, 120, 213, 226
transformational change 196
trends, scenario planning 167–73

value chain activities 69–71, 115–17, 210, 215, 226

vertical differentiation 63, 200
vertical integration 29
virtual enterprise structure 88–9
wage structures 220
weaknesses, SWOT analysis 75, 214
workers *see* operatives